Lessons in mathematics for the
Dyslexic Student & Visual Learner

Just the Facts!
Workbook

3

By Cheryl Orlassino

Check out our reading programs at www.BlastOffToReading.com

I Can Fly Reading Program for ages 5 to 7
Blast Off to Reading! for ages 7 to 13
A Workbook for Dyslexics, for ages 13+

Trademarks and trade names shown in this book are strictly for reading/decoding purposes and are the property of their respective owners. The author's references herein should not be regarded as affecting their validity.

Copyright © 2019 by Cheryl Orlassino. All rights reserved.
Some Images used in this book are licensed and
copyrighted by © GraphicsFactory.com

Published by Reading Specialists of Long Island, LLC
Centereach, New York

No part of this work may be reproduced or transmitted in any form or by any means, electronic or mechanical, including photocopying and recording, or by any information storage or retrieval system without the prior written permission of Cheryl Orlassino, unless such copying is permitted by federal copyright law.

e-mail all inquiries to:

math@help4dyslexia.com

For our math resources, go to:

www.help4dyslexia.com/math.html

Printed in the United States of America

First Edition - Rev A
International Standard Book Number (ISBN): 978-0-9831996-1-8

Table of Contents

Introduction..5
Lesson 1:   Skip Counting...6
Lesson 2:   Adding with Dots..12
Lesson 3:   Addition Double Facts...18
Lesson 4:   Multiplication..23
Lesson 5:   The Commutative Property..30
Lesson 6:   Multiplying by 2..37
Lesson 7:   Multiplying by 3, 5, and Multiples of 10................................43
Lesson 8:   Multiplying by 9..48
Lesson 9:   Multiplying by 11..54
Lesson 10: Multiplying by 4..59
Lesson 11: The Times Table Array..64
Lesson 12: Multiplying Double Facts...69
Lesson 13: Multiplying by 3 Revisited...76
Lesson 14: Multiplying 7x6, 7x8, and 8x6..82
Lesson 15: Multiplying by Zero..91
Lesson 16: Factors and Products...95
Lesson 17: Commas in Large Numbers..99
Lesson 18: Place Value of Numbers..105
Lesson 19: Multiplying Multiple Digits by Single Digits.........................109
Lesson 20: More on Multiplying Multiple Digits...................................114
Lesson 21: Multiplying Multiple Digits by 2 Digit Numbers....................118
Lesson 22: Multiplying Multiple Digits by 3 Digit Numbers....................123
Lesson 23: Greater, Less Than, and Equal..128
Lesson 24: Rounding Numbers...134
Lesson 25: Multiplying Multiples of Ten...139
Lesson 26: Estimating..144

Lesson 27: Expanding Numbers..149
Lesson 28: Multiplying with the Box Method.............................153
Lesson 29: The Distributive Property.......................................159
Lesson 30: Introduction to Exponents......................................164
Lesson 31: Division..169
Lesson 32: Word Problems..176
Lesson 33: Division - Practice & Review..................................181
Lesson 34: Fractions..185
Lesson 35: Fractions on a Number Line & Ruler.......................190
Lesson 36: Comparing Fractions..195
Lesson 37: Telling Time on an Analog Clock............................199
Lesson 38: Area of a Square & Rectangle................................204
Lesson 39: Perimeter of a Closed Shape.................................209
Lesson 40: Finding a Length, Given the Area...........................214
Lesson 41: Volume..219
Lesson 42: Mass...224
Lesson 43: Graphing Data..229
Lesson 44: Patterns..234
Lesson 45: Quadrilaterals..238
Answers..243

Introduction

This workbook is designed for the visual learner, in that, when able, illustrations are used to help the student gain understanding of the topic at hand. In addition, *Just the Facts* incorporates short stories (which includes a picture) for difficult to remember multiplication facts. It is recommended to use these stories as a quick ways to memorize certain math facts, however, it is still necessary for the student to understand the mathematical concepts.

The topics are arranged sequentially, so that lessons build on earlier lessons. Thus, **lessons should not be skipped and should be done in order.** In addition, concepts learned in prior lessons will often be revisited so that material taught will not be forgotten. This repetition is key in developing a good understanding and memorization of mathematical principles.

Instructions

Each lesson is broken up into two sections.

The first section is the lesson.
You must read the lesson to your student,
and then do all of the problems together.

The second part is the independent work,
which is 1 to 3 pages. This is for the student
to complete by him or herself.

Before you begin your next lesson,
review the independent work together, having the
student make the necessary corrections.

*We recommend periodically checking our website for
updates and additions to this program.*

Lesson 1: Skip Counting

Instead of counting by one's, we can **skip count**, to find the total number much quicker. For example, if you go on a trip with a large group, your leader will probably count everyone many times, to make sure everyone is together. To do this, the leader will probably count by *twos*, which takes half the time that it would take if he or she counted by ones.

Skip counting by 2

Start at 2, then draw the arrows to skip count every SECOND number.

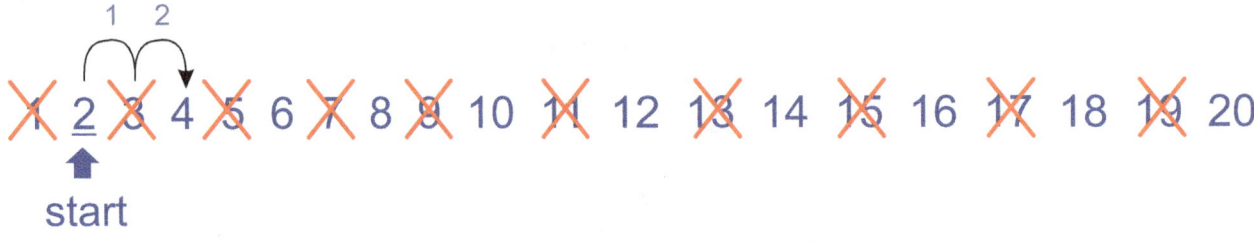

(1.1) Write the numbers: 2, ___, ___, ___, ___, ___, ___, ___, ___, ___

Skip counting by 3

Start at 3, then draw the arrows to skip count every THIRD number.

(1.2) Write the numbers: 3, ___, ___, ___, ___, ___

Skip counting by 5

Start at 5, then draw the arrows to skip count every FIFTH number.

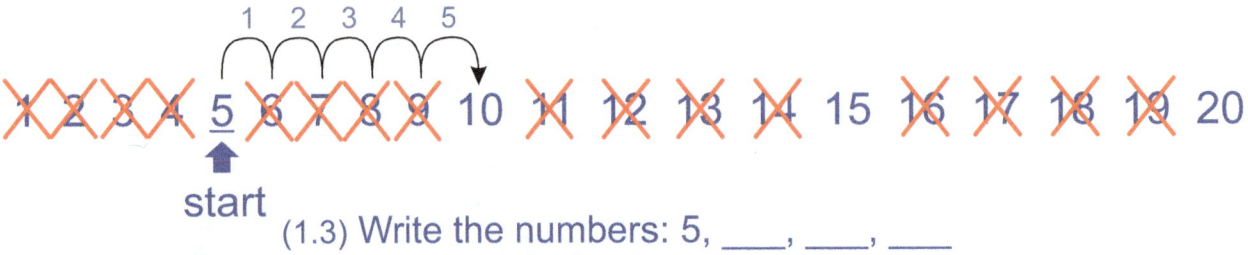

(1.3) Write the numbers: 5, ___, ___, ___

Just the Facts! Workbook

Lesson 1: Skip Counting

To skip count, add the number, that you are counting by, to the **previous** number (the number before the new number).

Below shows how to skip count by 10, which is also called "counting by tens".

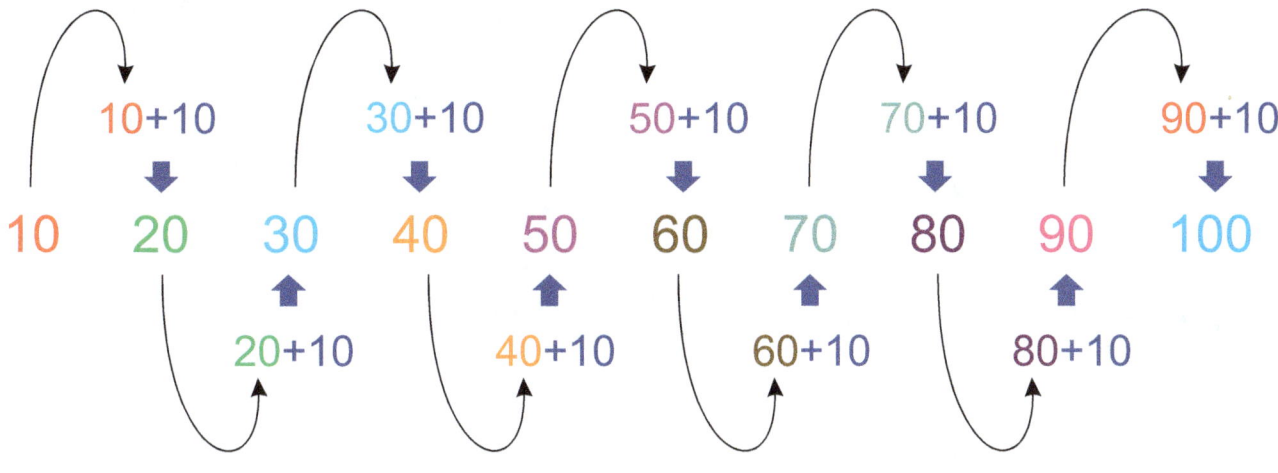

Skip count to fill in the missing numbers below.

(1.4)	(1.5)	(1.6)	(1.7)
2	3	5	10
4	6	10	20
8	12	20	40
10	15		50
14	21	35	70
16		40	
	27		
		50	100

Just the Facts! Workbook

Lesson 1: Skip Counting

Skip count to complete the sentences below.

(1.8)

There are _____ beans in all.

(1.9)

There are _____ fish in all.

(1.10)

There are _____ scoops in all.

(1.11) These are dimes, which are 10 cents each.

There are _____ cents in all.

(1.12) These are nickels, which are 5 cents each.

There are _____ cents in all.

(1.13)

There are _____ fish in all.

Just the Facts! Workbook

Lesson 1: Skip Counting

Skip count and **add** to complete the sentences below.

(1.14)

There are _____ beans in all.

(1.15)

There are _____ fish in all.

(1.16)

There are _____ scoops in all.

(1.17)
These are dimes and a nickel, worth 10 and 5 cents.

There are _____ cents in all.

(1.18)
These are nickels and pennies, which are 5 cents and 1 cent.

There are _____ cents in all.

(1.19)
These are a mix of dimes, nickels and pennies.

There are _____ cents in all.

Just the Facts! Workbook

Lesson 1: Skip Counting

Independent Work - Page 1

Skip count to complete the sentences below.

1.

   There are _____ fish in all.

2.

   There are _____ beans in all.

3.

   There are _____ cents in all.

4.

   There are _____ cents in all.

5.

   There are _____ scoops in all.

6.

   There are _____ fish in all.

7.

   There are _____ beans in all.

8.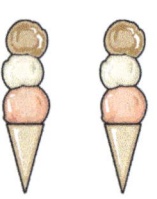

   There are _____ scoops in all.

Just the Facts! Workbook

Lesson 1: Skip Counting

Independent Work - Page 2

Skip count and **add** to complete the sentences.

9.

   There are _____ cents in all.

10.

    There are _____ cents in all.

11.

    There are _____ cents in all.

12.

    There are _____ cents in all.

13.

    There are _____ cents in all.

14.

    There are _____ cents in all.

15.

    There are _____ cents in all.

16.

    There are _____ cents in all.

Lesson 2: Adding with Dots

You've probably been adding numbers for a while now. We're often told to not use our fingers, however, when you can't remember a math fact, it is very helpful to count using something. Fingers are not a good choice, since you're often holding a pencil. You'd have to put the pencil down, count, then pick the pencil back up - very time consuming!

Using the five dot method is a good choice.
If you like to use tallies, that's good too.

⁙ ← Here's our five dots.

Here's tallies. ➡ 𝍩

Steps to Add Numbers With the Five Dot Method

1. Find the largest number in the equation. In this case, it's 5.

$$\begin{array}{r} 5 \\ +4 \\ \hline \end{array}$$

Start at 5, use 4 dots

2. The number that is being added to the larger number will be the number of dots that you will use. In this case, you will use 4 dots (draw the dots).

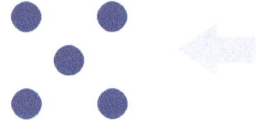

$$\begin{array}{r} 5 \\ +4 \\ \hline 9 \end{array}$$

6, 7, 8, 9

Stop at 9

3. Now, start at 5 (the largest number), and count up using the dots: 6 - 7 - 8 - 9. The answer is 9.

Just the Facts! Workbook

Lesson 2: Adding with Dots

5
+6

Start at 6, use 5 dots

7, 8, 9, 10, 11

6
+8

Start at 8, use 6 dots

9, 10, 11, 12, 13, 14

Write the number of dots you see below.
In the second row, start at five and then count up using the remaining dots.

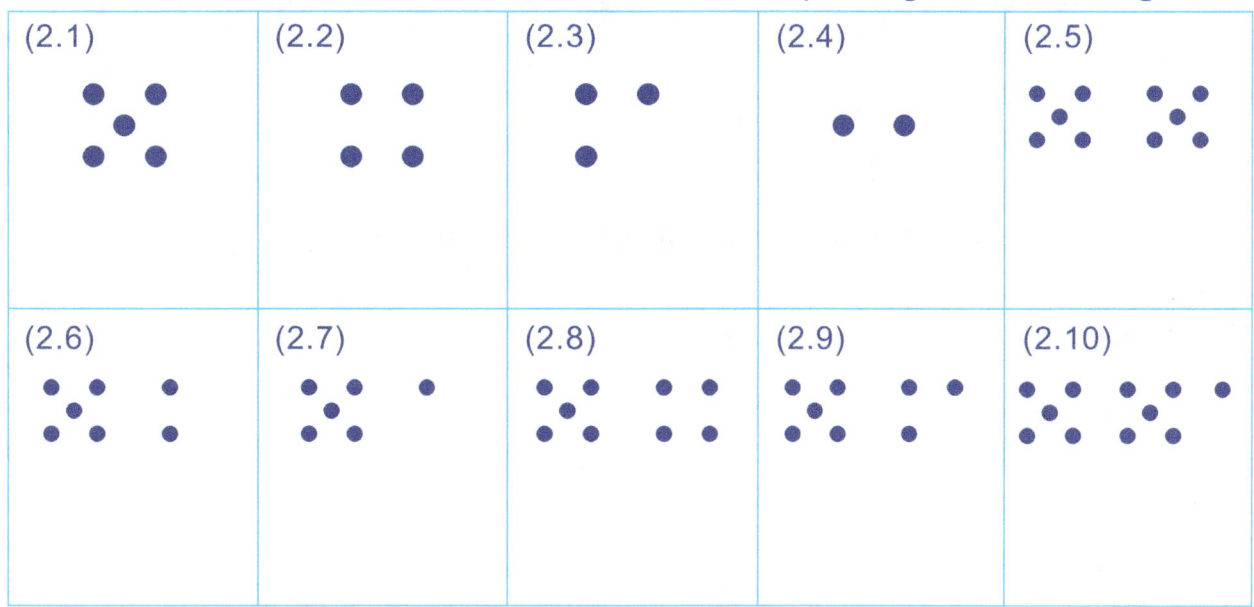

Use the five dot method to find the answers.

(2.11) 6 +4

(2.12) 5 +3

(2.13) 9 +3

(2.14) 9 +6

(2.15) 2 +8

(2.16) 7 +6

Just the Facts! Workbook

Lesson 2: Adding with Dots

For adding a **two** digit number to a **single** digit number you can also use the five dot method, or you can use the carry method, which you most likely have learned already.

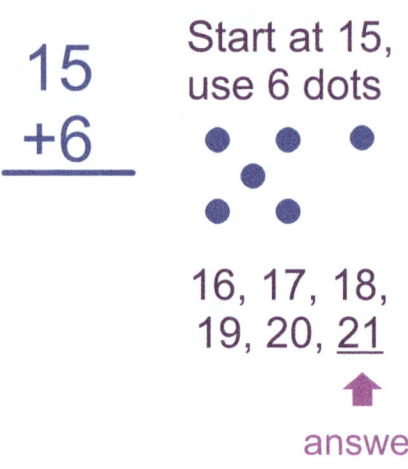

15
+6

Start at 15, use 6 dots

16, 17, 18, 19, 20, 21
↑
answer

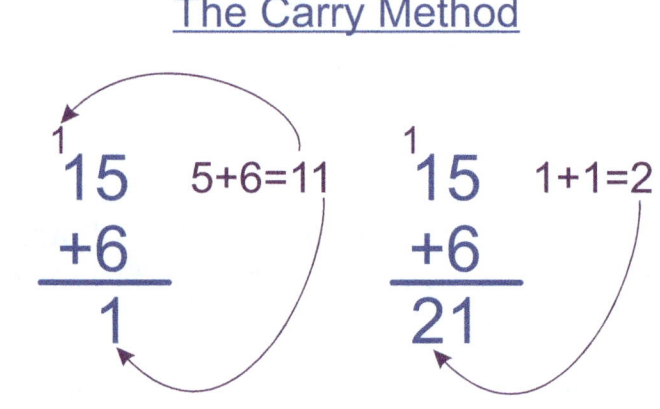

The Carry Method

Use the five dot method to find the answers.

(2.17) 16 +3	(2.18) 19 +4	(2.19) 28 +6
(2.20) 18 +5	(2.21) 12 +9	(2.22) 37 +6
(2.23) 29 +4	(2.24) 55 +7	(2.25) 46 +6
(2.26) 33 +7	(2.27) 88 +4	(2.28) 98 +5

Just the Facts! Workbook

Lesson 2: Adding with Dots

Count by 2's, 3's, or 5's to find the number of dots.

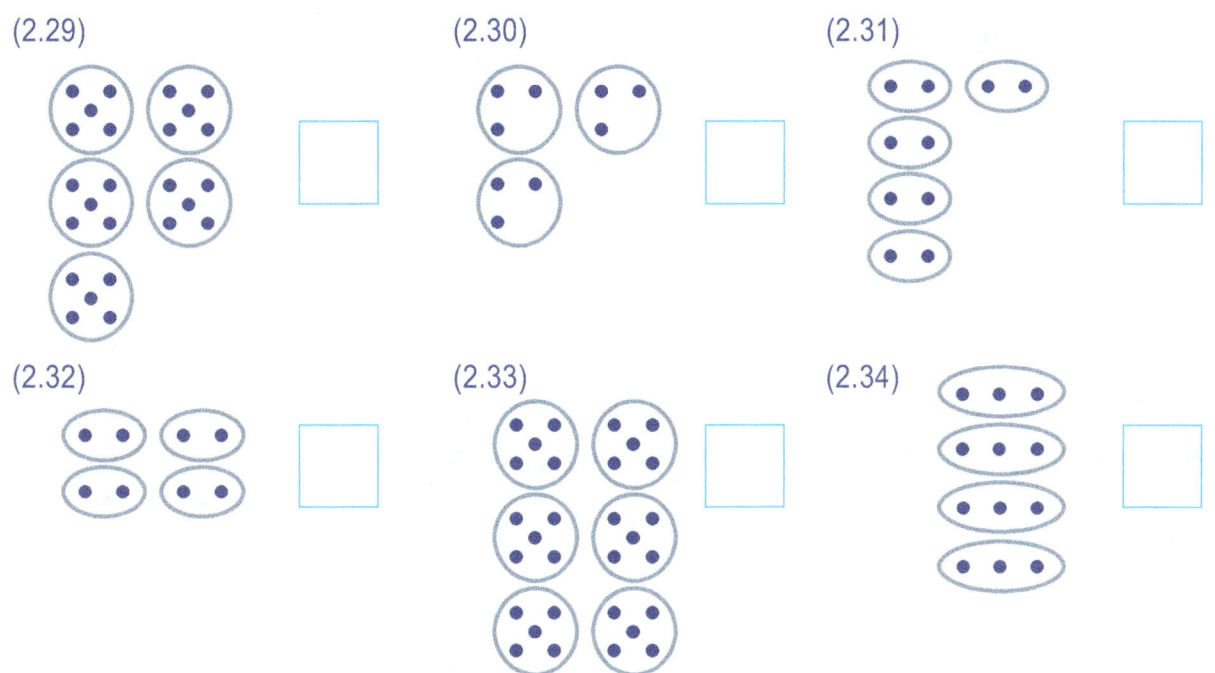

Write the number of dots. When you can, count by 5's (if it helps, first circle the groups of five).

(2.35)

(2.36)

(2.37)

(2.38)

(2.39)

(2.40)

Just the Facts! Workbook

Lesson 2: Adding with Dots

Independent Work - Page 1

Write the number of dots.

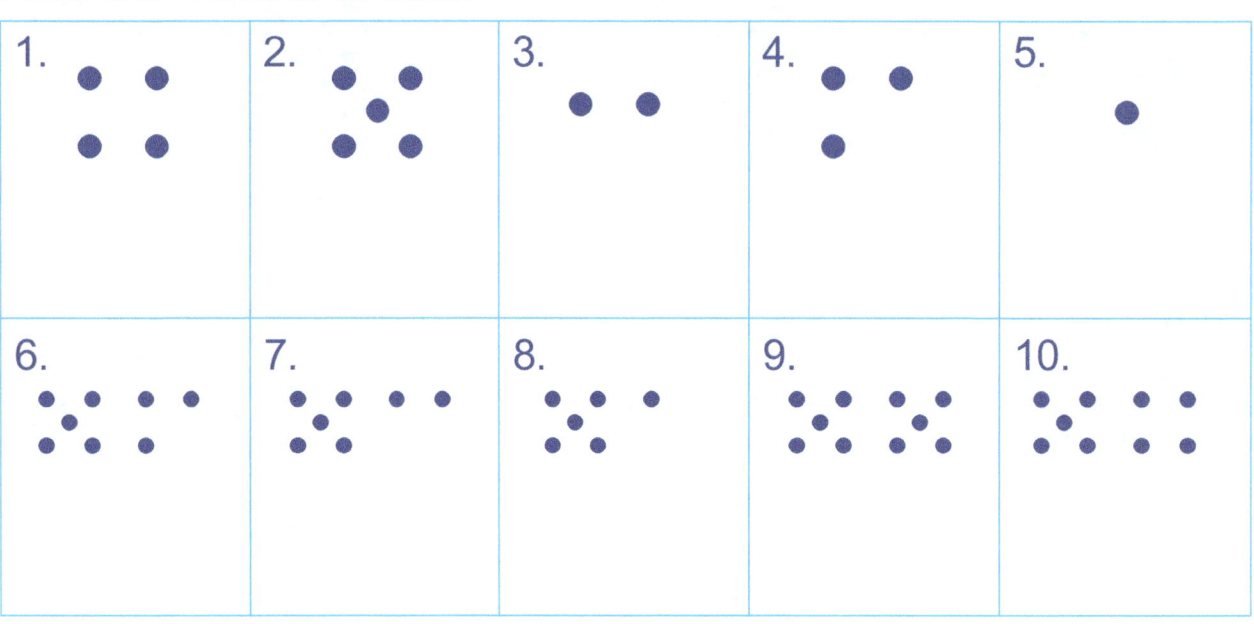

Use the five dot format to find the answers.

11. 6 +5	12. 9 +3	13. 8 +5
14. 7 +9	15. 7 +5	16. 8 +6
17. 29 +6	18. 12 +8	19. 17 +6

Just the Facts! Workbook

Lesson 2: Adding with Dots

Independent Work - Page 2

Count by 2's, 3's, or 5's to find the number of dots.

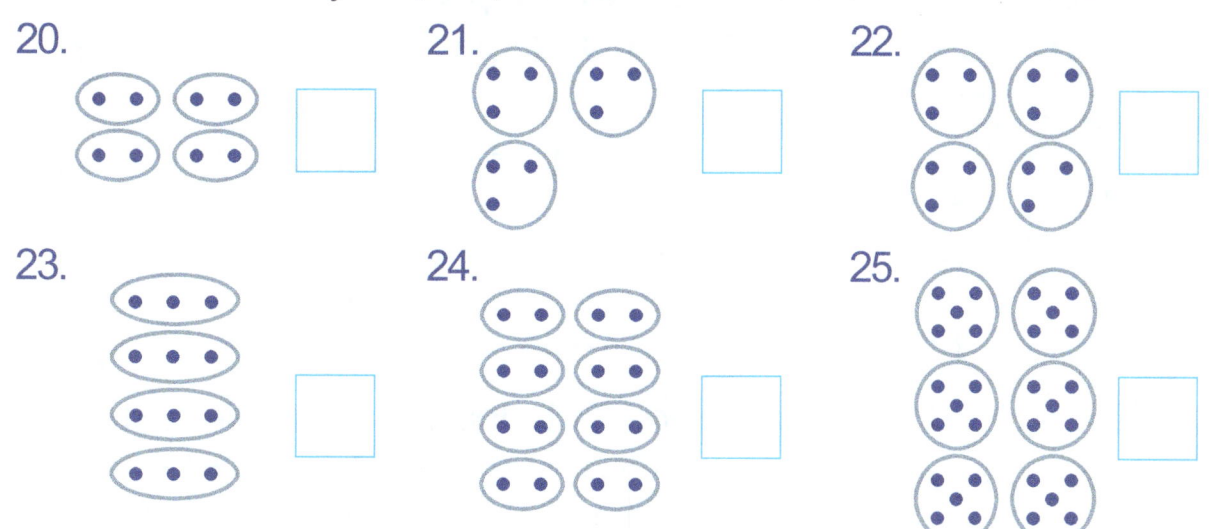

Write the number of dots. When you can, count by 5's (if it helps, first circle the groups of five).

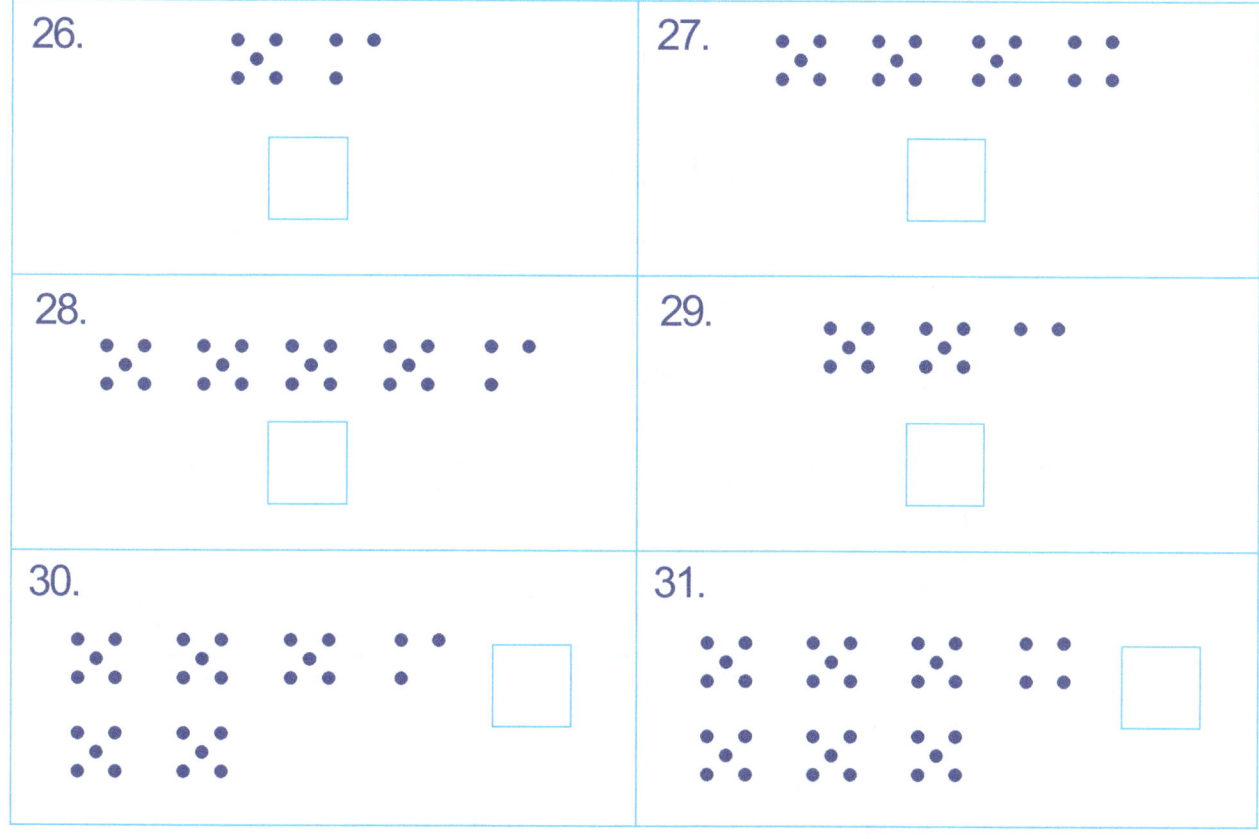

Just the Facts! Workbook

Lesson 3 — Addition Double Facts

Knowing your double facts will help you later on when you start to multiply.

1+1	2
2+2	4
3+3	6
4+4	8
5+5	10
6+6	12
7+7	14
8+8	16
9+9	18
10+10	20

Notice how the answers are the same as when you count by 2's!

The goal is to learn the double facts so that you know them by heart and don't have to use dots, or finger counting to get the answer.

* Put these on flash cards and review daily until mastered.

Double Facts Practice

Try to use your memory to find the answers below.

(3.1)   4
       +4

(3.2)   3
       +3

(3.3)   9
       +9

(3.4)   6
       +6

(3.5)   7
       +7

(3.6)   2
       +2

(3.7)   8
       +8

(3.8)   1
       +1

(3.9)   5
       +5

Just the Facts! Workbook

Lesson 3: Addition Double Facts

Skip count to fill in the missing numbers below.

(3.10)	(3.11)	(3.12)	(3.13)
2	3	5	10
4		10	20
8	12		
10	15		50
14	21	35	70
16	24	40	
	27		
		50	100

Use the five dot method to find the answers below.

(3.14) 9
       +4

(3.15) 8
       +6

(3.16) 5
       +3

(3.17) 9
       +8

(3.18) 8
       +4

(3.19) 9
       +6

(3.20) 17
       +8

(3.21) 19
       +7

(3.22) 16
       +5

Just the Facts! Workbook

Lesson 3: Addition Double Facts

Draw lines to match the double facts below.

(3.23) 3 + 3 =

(3.24) 7 + 7 =

(3.25) 2 + 2 =

(3.26) 5 + 5 =

(3.27) 1 + 1 =

(3.28) 9 + 9 =

(3.29) 4 + 4 =

(3.30) 6 + 6 =

(3.31) 8 + 8 =

2
4
6
8
10
12
14
16
18

Circle the correct answers for the equations below.

(3.32) 8 + 8 =   **12   16   9**   (3.36) 9 + 9 =   **16   14   18**

(3.33) 7 + 7 =   **14   12   18**   (3.37) 2 + 2 =   **6   8   4**

(3.34) 4 + 4 =   **2   16   8**   (3.38) 6 + 6 =   **16   10   12**

(3.35) 5 + 5 =   **12   10   14**   (3.39) 3 + 3 =   **12   6   8**

Just the Facts! Workbook

Lesson 3: Addition Double Facts

Independent Work - Page 1

Circle the correct answers for the equations below.

1. 6 + 6 =   14   12   10
2. 8 + 8 =   12   16   24
3. 2 + 2 =   4   8   2
4. 7 + 7 =   21   12   14

5. 4 + 4 =   16   8   7
6. 9 + 9 =   19   18   12
7. 5 + 5 =   6   11   10
8. 3 + 3 =   9   6   8

Skip count and **add** to complete the sentences below.

9.

There are _____ fish in all.

10.

There are _____ beans in all.

11.

There are _____ cents in all.

12.

There are _____ cents in all.

13.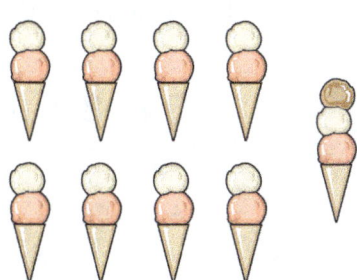

. There are _____ scoops in all.

14.

There are _____ cents in all.

Lesson 3: Addition Double Facts

Independent Work - Page 2

Draw lines to match the double facts below.

15. 7 + 7 =

16. 4 + 4 =

17. 2 + 2 =

18. 8 + 8 =

19. 3 + 3 =

20. 5 + 5 =

21. 1 + 1 =

22. 9 + 9 =

23. 6 + 6 =

2
4
6
8
10
12
14
16
18

Solve the equations below, try to use your memory.

24. 4 + 4 = _____   27. 7 + 7 = _____   30. 8 + 8 = _____

25. 3 + 3 = _____   28. 2 + 2 = _____   31. 6 + 6 = _____

26. 9 + 9 = _____   29. 5 + 5 = _____   32. 1 + 1 = _____

Just the Facts! Workbook

Lesson 4: Multiplication

Instead of adding the same numbers together, like 5+5+5+5, which can take a lot of time, we can **multiply** them. The multiplication symbol can look like the letter 'x', although sometimes it is written as a '*' or a dot. There are a lot of other ways to write these type of equations, but for now, we will use the 'x' as a symbol for multiplication.

> When we see a multiplication equation, we can think that the multiplication symbol means "groups of".
>
> 'x' means "groups of "

Below we have

4 groups of 5 dots, or

4 x 5

Below we have

6 groups of 5 dots, or

6 x 5

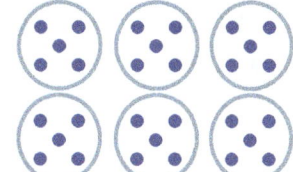

Below we have

6 groups of 4 dots, or

6 x 4

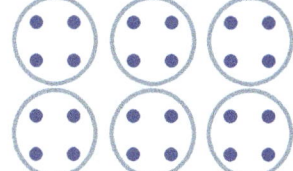

Below we have

4 groups of 3 dots, or

4 x 3

Just the Facts! Workbook

Lesson 4: Multiplication

Fill in the blanks to make an equation for the dots below.

(4.1)

___ groups of ___ dots

___ x ___

(4.2)

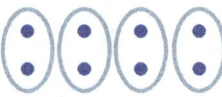

___ groups of ___ dots

___ x ___

(4.3)

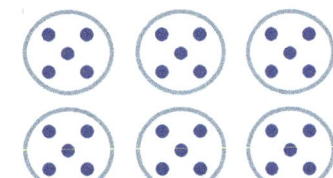

___ groups of ___ dots

___ x ___

(4.4)

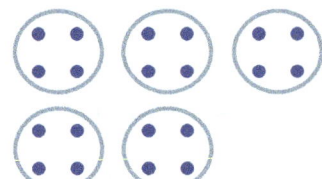

___ groups of ___ dots

___ x ___

(4.5)

___ groups of ___ dots

___ x ___

(4.6)

___ groups of ___ dots

___ x ___

Just the Facts! Workbook

Lesson 4: Multiplication

We can answer these multiplication problems by counting the dots, which can take a long time, or by skip counting, which is much faster.

4 groups of **5** dots, or

4 x 5 = **20**

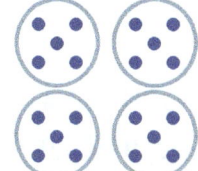

6 groups of **5** dots, or

6 x 5 = **30**

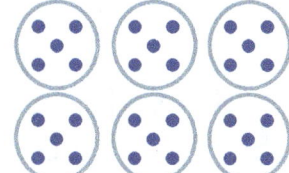

3 groups of **3** dots, or

3 x 3 = **9**

6 groups of **2** dots, or

6 x 2 = **12**

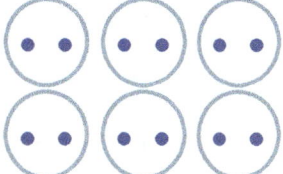

Sometimes we can't skip count, so we either have to have the answers memorized, or we will need another way to find the answer.
We will see more of these types of problems later, but for now, if you can't skip count, then you should **count** the dots.

4 groups of **6** dots, or

4 x 6 = **24**

Just the Facts! Workbook

Lesson 4: Multiplication

Word problems are very important in math. They are real world problems that show you how math is used in everyday life. When you have a word problem you have to create an equation using the information that is given to you.

Read the word problems below, underline the numbers, write the equations, and solve by skip counting.

(4.7)
Jim has four bags of beans. Each bag has five beans.
How many beans does Jim have altogether?
Draw the beans in each bag, then write out the equation and solve.
* You can solve by counting by fives.

___ groups of ___   ___ x ___ = ___

(4.8)
Jade has three bowls of fish. Each bowl has two fish.
How many fish does Jade have?
Draw the fish in each bowl, then write out the equation and solve.
* You can solve by counting by twos.

___ groups of ___   ___ x ___ = ___

Just the Facts! Workbook

Lesson 4: Multiplication

Independent Work - Page 1

Skip count to complete the tables below.

1.
2
4
8
10
14
20

2.
3
15
21
27

3.
5
10
40
50

4.
10
20
100

Double facts practice; try to use your memory to find the answers.

5. 4
   +4

6. 3
   +3

7. 9
   +9

8. 6
   +6

9. 7
   +7

10. 2
    +2

11. 8
    +8

12. 1
    +1

13. 5
    +5

Just the Facts! Workbook

Lesson 4: Multiplication

Independent Work - Page 2

Fill in the blanks to make an equation for the dots below, and then skip count to get the answers.

14.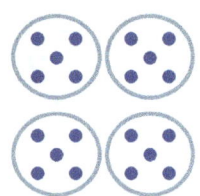

___ groups of ___ dots

___ x ___ = ___

15.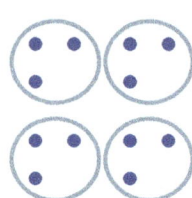

___ groups of ___ dots

___ x ___ = ___

16.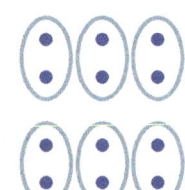

___ groups of ___ dots

___ x ___ = ___

17.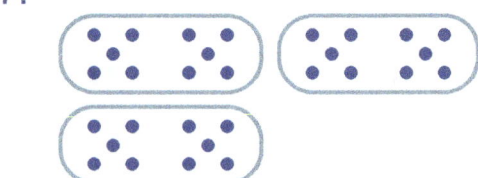

___ groups of ___ dots

___ x ___ = ___

18.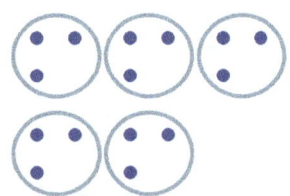

___ groups of ___ dots

___ x ___ = ___

19.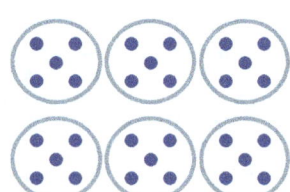

___ groups of ___ dots

___ x ___ = ___

Just the Facts! Workbook

Lesson 4: Multiplication

Independent Work - Page 3

Read the word problems below, write out the equations, and solve.

20.

Jim had five bags of beans. Each bag had 5 beans. How many beans did Jim have altogether?
Draw the beans in each bag, then write out the equation.

___ groups of ___   ___ x ___ = ___

21.

Jade had four bowls of fish. Each bowl had five fish. How many fish did Jade have?
First, draw the fish in each bowl, then write out the equation.

___ groups of ___   ___ x ___ = ___

22.

There are six faces. Each face has two eyes. How many eyes are there altogether?

___ groups of ___   ___ x ___ = ___

Lesson 5: The Commutative Property

Before we move on, there's a math property you need to know.
It's called the **commutative property**.
This property tells us is that the numbers in an addition or multiplication equation can be switched around.
Pretty simple!

Commutative Property of **Addition**

2 + 1 = <u>3</u>   or   1 + 2 = <u>3</u>

Commutative Property of **Multiplication**

2 x 6 = <u>12</u>   or   6 x 2 = <u>12</u>

Notice how the answers are the same, even though the numbers switched places.

Fill in the blanks below using the commutative property.

(5.1) 6 + 8 = 8 + ____   (5.5) 9 x 5 = 5 x ____

(5.2) 5 + 3 = 3 + ____   (5.6) 4 x 8 = 8 x ____

(5.3) 4 + 9 = 9 + ____   (5.7) 5 x 3 = 3 x ____

(5.4) 7 + 2 = 2 + ____   (5.8) 8 x 6 = 6 x ____

Just the Facts! Workbook

Lesson 5: The Commutative Property

Circle the trophies below in groups of 2.

(5.9) There are _____ groups of 2, or ___ x 2

Circle the trophies blow in groups of 8.

(5.10) There are _____ groups of 8, or ___ x 8

Look at the two groups of trophies above. Are they equal?
8 groups of 2 is the **same** as 2 groups of 8, or

8 x 2 = 2 x 8

(5.11)    (5.12)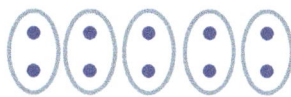

____ groups of ____   ____ groups of ____

___ x ___ = ___   ___ x ___ = ___

(5.13) Do the two groups above have the same number of dots?

Lesson 5: The Commutative Property

You can use the **commutative property** to help multiply. For example, multiplying by six is difficult, and memorizing the facts for the six times tables is also difficult. If you can, switch the numbers around so it becomes easier.

5 x 6 = **Five** groups of **six.**

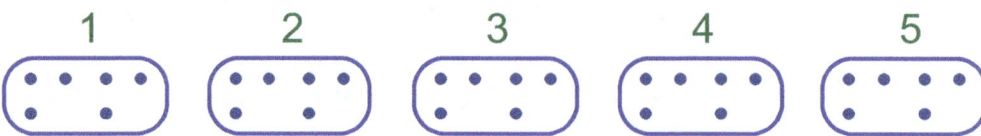

Switch the numbers around.

6 x 5 = **Six** groups of **five.**

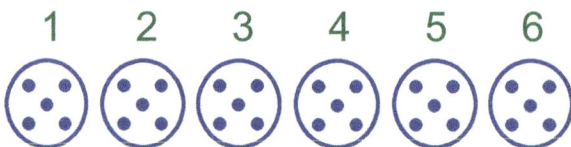

Now you can count by 5's.

Use the **commutative property** to rewrite the equations, and then solve by skip counting.

(5.14) 4 x 5 = ___ x ___ = ___   (5.18) 4 x 10 = ___ x ___ = ___

(5.15) 7 x 5 = ___ x ___ = ___   (5.19) 6 x 3 = ___ x ___ = ___

(5.16) 4 x 3 = ___ x ___ = ___   (5.20) 7 x 2 = ___ x ___ = ___

(5.17) 6 x 2 = ___ x ___ = ___   (5.21) 7 x 10 = ___ x ___ = ___

The word "**commute**" means **to travel.**
The numbers can **travel** from one spot to another!

Lesson 5: The Commutative Property

Fill in the blanks, to write the equation on the left, then use the commutative property to rewrite the equation on the right so that you can solve by skip counting.

(5.22)

___ groups of ___ dots   ___ groups of ___ dots

___ x ___ = ___   ___ x ___ = ___

(5.23)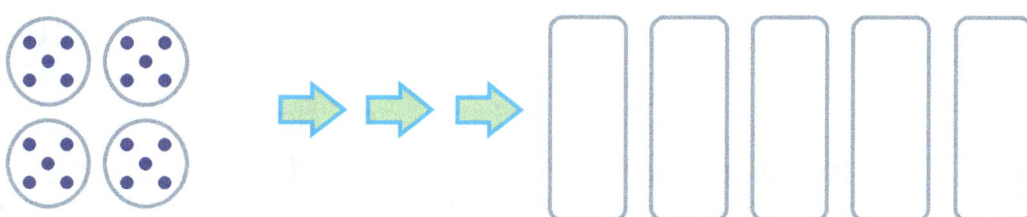

___ groups of ___ dots   ___ groups of ___ dots

___ x ___ = ___   ___ x ___ = ___

(5.24) Skip count by 5 and 10, fill in the blanks.

5, ___, ___, ___, ___, ___, ___, ___, ___, ___

10, ___, ___, ___, ___, ___, ___, ___, ___, ___

Just the Facts! Workbook

Lesson 5: The Commutative Property

Independent Work - Page 1

Skip count to complete the tables below.

1.	2.	3.	4.
2	3	5	10
	15		
	21		
	27		
20		50	100

Double facts practice;
try to use your memory to find the answers.

5.   8
    +8
	6. 5
	    +5
		7. 9
		    +9

8.   3
    +3
	9. 4
	    +4
		10. 2
		    +2

11.   1
    +1
	12. 7
	    +7
		13. 6
		    +6

Just the Facts! Workbook

Lesson 5: The Commutative Property

Independent Work - Page 2

Fill in the blanks, to write the equation on the left, then use the **commutative** property to rewrite the equation on the right so that you can solve by skip counting.

14.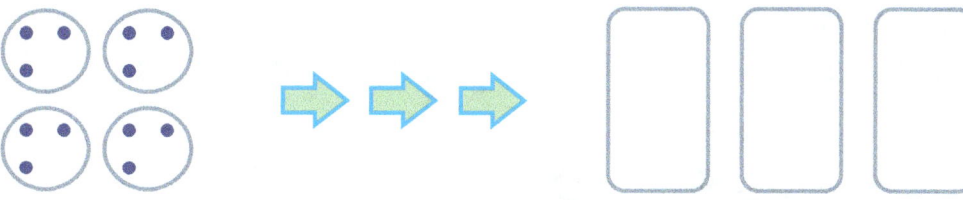

___ groups of ___ dots   ___ groups of ___ dots

___ x ___ = ___   ___ x ___ = ___

15.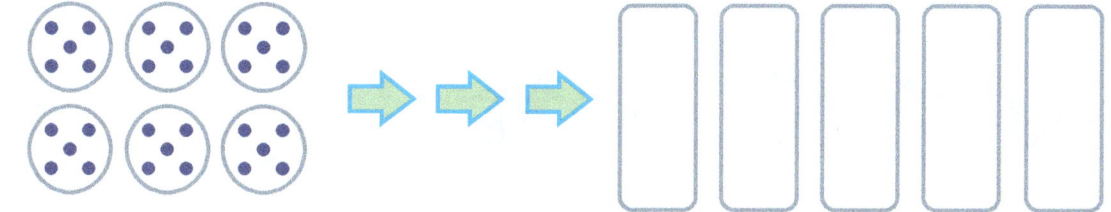

___ groups of ___ dots   ___ groups of ___ dots

___ x ___ = ___   ___ x ___ = ___

16.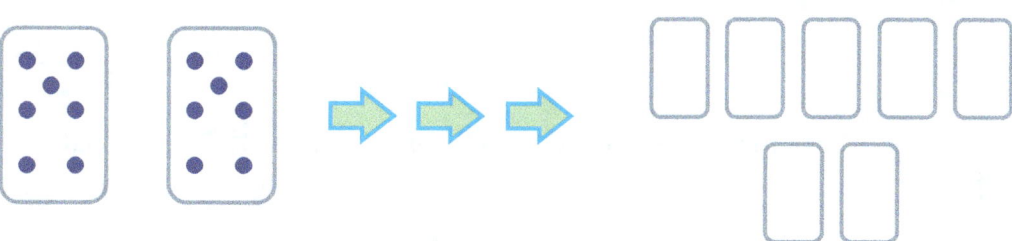

___ groups of ___ dots   ___ groups of ___ dots

___ x ___ = ___   ___ x ___ = ___

Lesson 5: The Commutative Property

Independent Work - Page 3

Solve the word problems below. Remember, underline the numbers, then write out two equations (use the commutative property) and solve.

17. Jack is building three houses at one time. Each house needs ten doors. How many doors does Jack need to buy?

       ___ x ___ or ___ x ___

    Jack needs _____ doors.

18. Jack is building five houses at one time. Each house needs three doors. How many doors does Jack need to buy?

    ___ x ___ or ___ x ___

    Jack needs _____ doors.

19. Jack is building four houses at one time. Each house needs three doors. How many doors does Jack need to buy?

       ___ x ___ or ___ x ___

    Jack needs _____ doors.

20. Jack is building six houses at one time. Each house needs five doors. How many doors does Jack need to buy?

       ___ x ___ or ___ x ___

    Jack needs _____ doors.

Lesson 6

Multiplying by 2

As we saw earlier, to multiply by two, you can skip count, which can take a long time to do. If you know your addition double facts, such as 8+8, then you can do these problems faster.

1. Circle the above trophies in groups of 2's.

2. Now skip count to find the total.

(6.1) The total is _____.

3. Now write the equation to find the total, using **multiplication**.

(6.2) ____ groups of ____ trophies

____ x ____

4. Using the **commutative property** we can write the equation as

2 x 8

5. We now **rewrite** the equation using **an addition double fact**, and, since we know the double facts, we can solve it.

2 x 8 = **8 + 8** = 16

Just the Facts! Workbook

Lesson 6: Multiplying by 2

Fill in the blanks to make an equation for the dots below.

(6.3)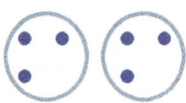

__2__ groups of __3__ dots

__2__ x __3__ or __3__ + __3__

= _____ dots

(6.4)

___ groups of ___ dots

___ x ___ or ___ + ___

= _____ dots

(6.5)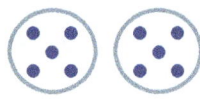

___ groups of ___ dots

___ x ___ or ___ + ___

= _____ dots

(6.6)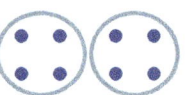

___ groups of ___ dots

___ x ___ or ___ + ___

= _____ dots

(6.7)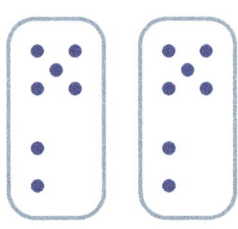

___ groups of ___ dots

___ x ___ or ___ + ___

= _____ dots

(6.8)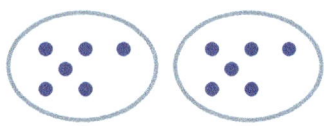

___ groups of ___ dots

___ x ___ or ___ + ___

= _____ dots

Lesson 6: Multiplying by 2

> Addition double facts, such 8+8, are important to know since you will use these when you multiply by 2.

Below we change the 2 times equations into addition double facts.

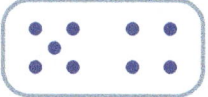 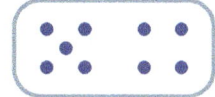

(6.9) 2 groups of 9 = 2 x 9 = __9__ + __9__ = _____

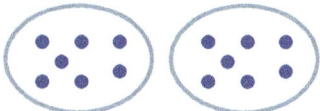

(6.10) 2 groups of 7 = 2 x 7 = _____ + _____ = _____

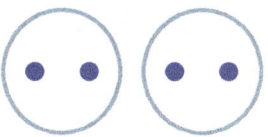

(6.11) 2 groups of 2 = 2 x 2 = _____ + _____ = _____

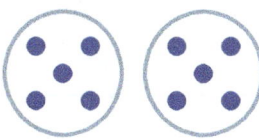

(6.12) 2 groups of 5 = 2 x 5 = _____ + _____ = _____

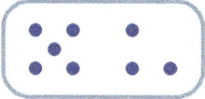 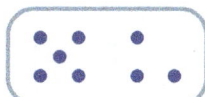

(6.13) 2 groups of 8 = 2 x 8 = _____ + _____ = _____

Just the Facts! Workbook

Lesson 6: Multiplying by 2

> Anytime you multiply by 2, you can use **addition double facts**.

Review: solve the addition double facts below.

(6.14) 2 + 2 = _____   (6.18) 6 + 6 = _____

(6.15) 3 + 3 = _____   (6.19) 7 + 7 = _____

(6.16) 4 + 4 = _____   (6.20) 8 + 8 = _____

(6.17) 5 + 5 = _____   (6.21) 9 + 9 = _____

Change the 2 times multiplication facts below into addition double facts and solve.

(6.22) 2 x 3 = __3 + 3__ = __6__

(6.23) 2 x 6 = _____ = _____

(6.24) 2 x 7 = _____ = _____

> If you forget your addition double facts, you can always count by twos.
>
> 2 x 4 = 2, 4, 6, **8**
>
> Keep in mind that it's faster if you use your double facts.

Just the Facts! Workbook

Lesson 6: Multiplying by 2

Independent Work - Page 1

Rewrite the problems using **addition double facts** and solve.

1. 2 x 4 = ___ + ___ = ___

2. 2 x 9 = ___ + ___ = ___

3. 2 x 5 = ___ + ___ = ___

4. 2 x 3 = ___ + ___ = ___

Rewrite the problems using **multiplication** and solve.

5. 6 + 6 = ___ x ___ = ___

6. 8 + 8 = ___ x ___ = ___

7. 2 + 2 = ___ x ___ = ___

8. 7 + 7 = ___ x ___ = ___

Write the multiplication equations and solve.

9.  2 groups of 4 = ___ x ___ = ___

10.  2 groups of 6 = ___ x ___ = ___

11.  2 groups of 8 = ___ x ___ = ___

Just the Facts! Workbook

Lesson 6: Multiplying by 2

Independent Work - Page 2

Circle the equations that are equal to the equations on the left (there may be more than one).

12.	8 x 2	8 + 2	8 + 8	2 x 8
13.	2 x 6	6 x 2	2 + 6	6 + 6
14.	5 x 2	5 + 5	2 x 5	2 + 5
15.	9 x 2	9 + 9	2 x 9	9 + 2

Create multiplication equations for the pictures below and solve.

16.  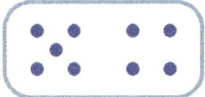   ___ X ___ = ___

17. 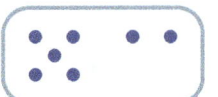 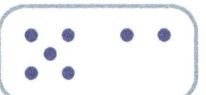   ___ X ___ = ___

18. 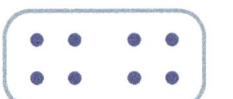 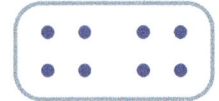   ___ X ___ = ___

19.    ___ X ___ = ___

Lesson 7: Multiplying by 3, 5, and Multiples of 10

Remember all of the skip counting exercises that you've done? That will help you when multiplying!

(7.1)

Here we have
<u>4</u> groups of <u>3</u> fish or

4 x 3

3 + 3 + 3 + 3 = ____

(7.2) Here we have <u>3</u> groups of <u>5</u> fish, or 3 x 5.

5 + 5 + 5 = ____

Note that we'd rather count by fives, than count by threes.

(7.3) To multiply by 10, you can either skip count by 10 or simply **add a zero** to the group number.

____ groups of 10 cents or ____ x ____

10 + 10 + 10 + 10 + 10

or add a zero after the 5: **5<u>0</u>**

Just the Facts! Workbook

Lesson 7: Multiplying by 3, 5, and Multiples of 10

Solve the problems below by adding zero to the number that 10 is multiplied by.

10 x 3 = 30   add a zero here

(7.4) 10 x 6 = _____

(7.7) 8 x 10 = _____

(7.5) 10 x 4 = _____

(7.8) 10 x 9 = _____

(7.6) 5 x 10 = _____

(7.9) 10 x 2 = _____

(7.10) 10 x 75 = _____

(7.13) 10 x 15 = _____

(7.11) 11 x 10 = _____

(7.14) 10 x 12 = _____

(7.12) 53 x 10 = _____

(7.15) 25 x 10 = _____

First, underline the number that you want to skip count by, and then solve by skip counting.

(7.16) 5 x 4 = 20

(7.19) 3 x 4 = _____

(7.17) 3 x 5 = _____

(7.20) 6 x 3 = _____

(7.18) 5 x 7 = _____

(7.21) 10 x 3 = _____

Lesson 7: Multiplying by 3, 5, and Multiples of 10

To multiply by any multiple of 10, such as 100, 1000, 10,000, etc., simply count the zeros and add that many zeros to the number you are multiplying by.

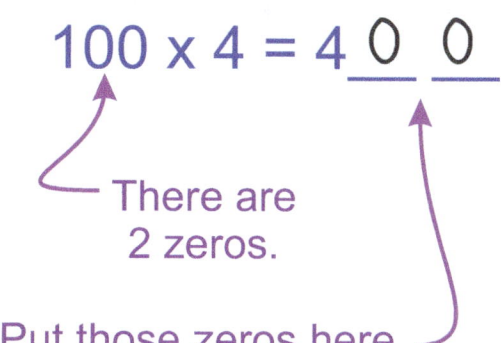

100 x 4 = 4 _0_ _0_

— There are 2 zeros.

Put those zeros here.

(7.22) 10 x 5 = 5 ___

(7.23) 100 x 5 = 5 ___ ___

(7.24) 1,000 x 5 = 5, ___ ___ ___

(7.25) 10,000 x 5 = 5 ___ , ___ ___ ___

After every three numbers from the right a comma is added. We will see more on this in a later lesson.

Solve the equations below; notice that commas are used after every **three** numbers from the **right** side of the number.

(7.26) 66 x 1,000 = ___ ___ , ___ ___ ___

(7.27) 951 x 100 = ___ ___ , ___ ___ ___

(7.28) 6,457 x 10 = ___ ___ , ___ ___ ___

(7.29) 32 x 1,000 = ___ ___ , ___ ___ ___

(7.30) 320 x 1,000 = ___ ___ ___ , ___ ___ ___

Lesson 7: Multiplying by 3, 5, and Multiples of 10

Independent Work - Page 1

	number of zeros	answer
1. 100 x 7 =		
2. 10 x 54 =		
3. 1,000 x 30 =		
4. 1000 x 400 =		

Circle the equations that are equal to the equations on the left (there may be more than one that is equal).

5.	12 x 2	2 x 12	12 + 12	12 + 2
6.	3 x 6	6 x 3	6 + 6 + 6	6 + 3
7.	6 + 6	2 x 6	6 x 2	6 x 6
8.	4 x 3	3 + 3 + 3 + 3	3 x 4	4 + 4 + 4
9.	3 x 2	2 + 2	2 + 2 + 2	3 + 3

Lesson 7: Multiplying by 3, 5, and Multiples of 10

Independent Work - Page 2

Use double addition facts to solve the problems below.

10. 2
    x 4

11. 2
    x 6

12. 2
    x 3

Count by fives to solve the problems below.

13. 5
    x 4

14. 5
    x 7

15. 5
    x 8

Count by threes to solve the problems below.

16. 3
    x 3

17. 3
    x 6

18. 3
    x 4

Solve the problems below, you can count by tens or simply add one zero to the number being multiplied.

19. 10
    x 8

20. 10
    x 4

21. 10
    x 6

Lesson 8: Multiplying by 9

There are three ways to find a 9 times fact.
1. Sheer memorization.
2. The hand method.
3. The two step number method.

Sheer memorization, is difficult, and, if you don't use it often, it's usually forgotten. Therefore, we will concentrate on the hand method and the two step number method.

The Hand Method

Each finger gets a number associated with it (see below).

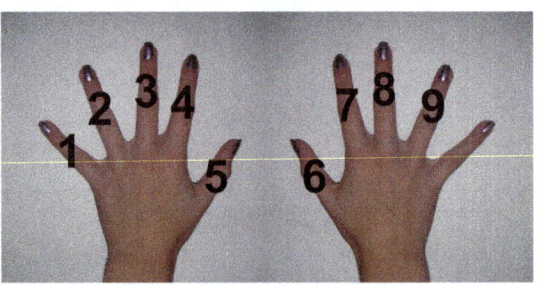

When you multiply 9 by a number,
that numbered finger goes down.
The answer is the two digit number that is
created by the fingers that are still up.
The first digit, of your answer, is to the left of the finger
that's down, and the second digit, of your answer, is
to the right of the finger that's down.

9 x 3

9 x 3 = 27

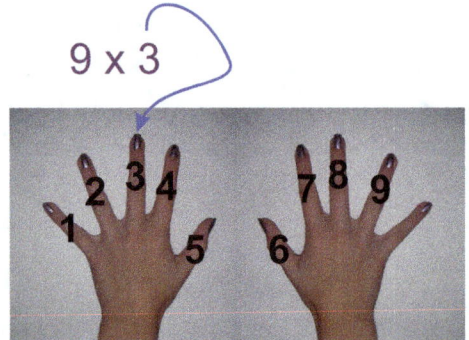

2 up   7 up
27

Just the Facts! Workbook

Lesson 8: Multiplying by 9

Below shows all the ways to multiply a number,
under ten, by nine.

1 x 9 = 9
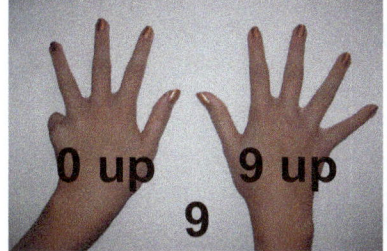
1st finger is down

2 x 9 = 18
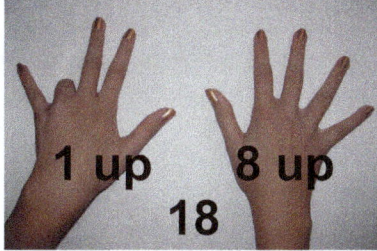
2nd finger is down

3 x 9 = 27
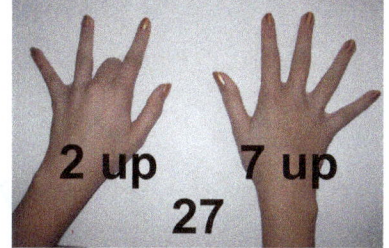
3rd finger is down

4 x 9 = 36
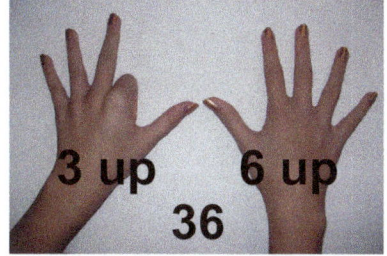
4th finger is down

5 x 9 = 45
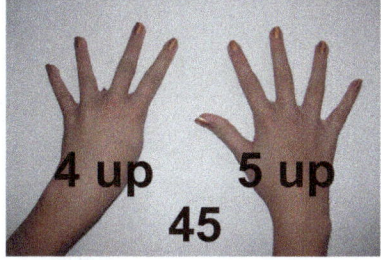
5th finger is down

6 x 9 = 54
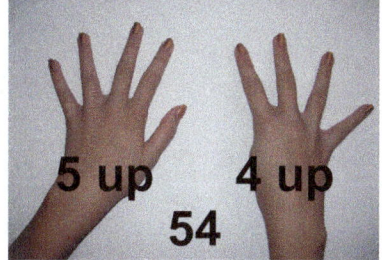
6th finger is down

7 x 9 = 63
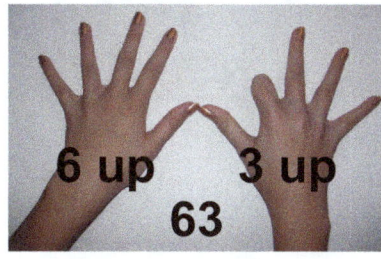
7th finger is down

8 x 9 = 72
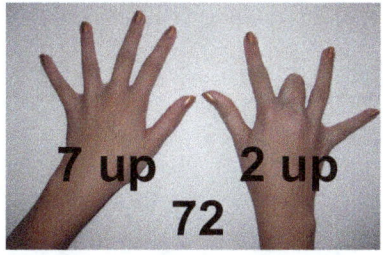
8th finger is down

9 x 9 = 81
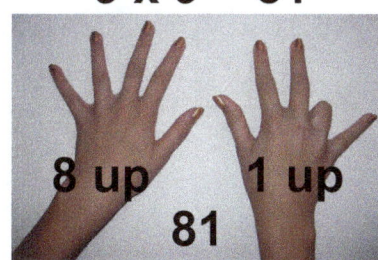
9th finger is down

Lesson 8: Multiplying by 9

Now, cover up the pictures on the right side and see if you can solve the equation by using your hands.

(8.1) 9 x 4 = _____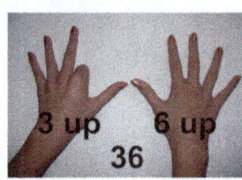

(8.2) 9 x 3 = _____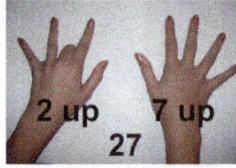

(8.3) 9 x 6 = _____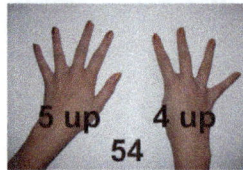

(8.4) 9 x 9 = _____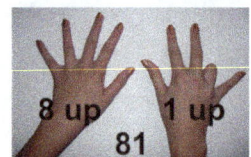

Solve the equations below.

(8.5) 9   (8.6) 9   (8.7) 9   (8.8) 9
     x 4   x 6   x 3   x 7

(8.9)   (8.10)   (8.11)   (8.12)
      5   8   2   9
    x 9   x 9   x 9   x 9

Just the Facts! Workbook   - 50 -

Lesson 8: Multiplying by 9

5 x 9 = ?   For this problem, you can use your hands, or you can count by fives.

9 x 9 = ?   For this problem, you can use your hands, or you can use the relationship to the addition problem by having the digits switch places; 18 to 81.

9 + 9 = 18
81

The Two Step Number Method

Step 1:
Subtract 1 from the number that isn't 9.

Step 2:
Subtract that number (from step 2) **from** 9.

9 x <u>3</u> =

3 - 1 = **<u>2</u>**

<u>2</u>

9 - 2 = **<u>7</u>**

2**<u>7</u>**

9 x <u>8</u> =	9 x <u>6</u> =	9 x <u>7</u> =
8 - 1 = **7**	6 - 1 = **5**	7 - 1 = **6**
9 - 7 = **2**	9 - 5 = **4**	9 - 6 = **3**
72	54	63

Use the hand method to check the answers.

Just the Facts! Workbook

Lesson 8: Multiplying by 9

Independent Work - Page 1

Use the hand method to solve the problems below.

1. 9
    x 4

2. 9
    x 6

3. 9
    x 3

4. 9
    x 7

5. 9
    x 9

6. 9
    x 2

Count the zeros to find the answers below.

7. 10
    x 4

8. 100
    x 9

9. 1,000
    x 3

10. 54
    x 10

11. 761
    x 10

12. 100
    x 23

Count by 5's to find the answers below.

13. 5 x 7 = _____

14. 3 x 5 = _____

15. 5 x 5 = _____

16. 5 x 8 = _____

17. 4 x 5 = _____

18. 5 x 6 = _____

Just the Facts! Workbook

Lesson 8: Multiplying by 9

Independent Work - Page 2

19. Jim spends $4.00 a day on lunch.
How much money does he need for 9 days?

___ x ___ = _____ dollars

20. Jim has $6.00 and then he finds $9.00 in his pocket.
How much money does he have now?

___ + ___ = _____ dollars

21. Jim reads three books a week.
How many books does Jim read in five weeks?

___ x ___ = _____ books

22. Jane bought nine books for $6.00 each.
How much did Jane spend on books?

___ x ___ = _____ dollars

23. Jane earns $8.00 an hour, and worked for nine hours.
How much money did Jane earn?

___ x ___ = _____ dollars

Lesson 9: Multiplying by 1 and 11

Multiplying by one may be obvious when you see it in terms of groups as done below.

> Remember, 'x' means "group(s) of"

Below we have

1 <u>group</u> <u>of</u> **5** dots, or

1 x 5

You can see that all you have is 5 dots.

1 <u>group</u> <u>of</u> **4** beans, or

1 x 4 = __4__

1 <u>group</u> <u>of</u> **3** fish, or

1 x 3 = __3__

1 <u>group</u> <u>of</u> **5** ice-cream cones

or 1 x 5 = __5__

1 <u>group</u> <u>of</u> **6** cents, or

1 x 6 = __6__

Lesson 9: Multiplying by 1 and 11

Do the problems below.

(9.1) 1 × 7 (9.2) 4 × 1 (9.3) 1 × 8 (9.4) 14 × 1 (9.5) 26 × 1

Multiplying by eleven is easy to remember,
but first lets look at it in terms of groups.

Below we have

2 group of 11 dots, or

2 x 11

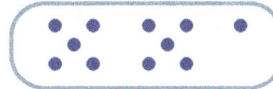 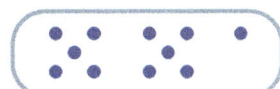

You can count the dots to see that we have 22 dots total.

2 x 11 = 22

Find the pattern and then finish the problems.

1 x 11 = 11   (9.6) 6 x 11 = ____

2 x 11 = 22   (9.7) 7 x 11 = ____

3 x 11 = 33   (9.8) 8 x 11 = ____

4 x 11 = 44   (9.9) 9 x 11 = ____

5 x 11 = 55

When multiplying by 11, simply write the digit being multiplied twice.

Lesson 9: Multiplying by 1 and 11

Review: addition double fact practice.

(9.10) 9 + 9 = _____   (9.13) 6 + 6 = _____   (9.16) 1 + 1 = _____

(9.11) 5 + 5 = _____   (9.14) 4 + 4 = _____   (9.17) 8 + 8 = _____

(9.12) 3 + 3 = _____   (9.15) 7 + 7 = _____   (9.18) 2 + 2 = _____

Review: skip count to complete the series of numbers.

(9.19)   3, _____, _____, _____, _____, _____, _____

(9.20)   5, _____, _____, _____, _____, _____, _____

Review: skip count, or use the addition double facts to solve.

(9.21) 9 x 2 = _____   (9.24) 2 x 7 = _____   (9.27) 8 x 5 = _____

(9.22) 5 x 4 = _____   (9.25) 3 x 4 = _____   (9.28) 8 x 2 = _____

(9.23) 3 x 3 = _____   (9.26) 6 x 5 = _____   (9.29) 2 x 2 = _____

Review: count the zeros to find the answers.

(9.30) 709 x 10 = _____

(9.31) 100 x 25 = _____

(9.32) 56 x 1,000 = _____

(9.33) 20 x 100 = _____

Just the Facts! Workbook

Lesson 9: Multiplying by 1 and 11

Independent Work - Page 1

Do the problems below.

1. $\begin{array}{r} 2 \\ \times\ 4 \\ \hline \end{array}$

2. $\begin{array}{r} 2 \\ \times\ 6 \\ \hline \end{array}$

3. $\begin{array}{r} 2 \\ \times\ 3 \\ \hline \end{array}$

4. $\begin{array}{r} 2 \\ \times\ 7 \\ \hline \end{array}$

5. $\begin{array}{r} 2 \\ \times\ 9 \\ \hline \end{array}$

6. $\begin{array}{r} 2 \\ \times\ 6 \\ \hline \end{array}$

7. $\begin{array}{r} 5 \\ \times\ 4 \\ \hline \end{array}$

8. $\begin{array}{r} 5 \\ \times\ 6 \\ \hline \end{array}$

9. $\begin{array}{r} 5 \\ \times\ 8 \\ \hline \end{array}$

10. $\begin{array}{r} 4 \\ \times\ 3 \\ \hline \end{array}$

11. $\begin{array}{r} 3 \\ \times\ 3 \\ \hline \end{array}$

12. $\begin{array}{r} 3 \\ \times\ 6 \\ \hline \end{array}$

13. $\begin{array}{r} 9 \\ \times\ 3 \\ \hline \end{array}$

14. $\begin{array}{r} 9 \\ \times\ 4 \\ \hline \end{array}$

15. $\begin{array}{r} 9 \\ \times\ 6 \\ \hline \end{array}$

16. $\begin{array}{r} 100 \\ \times\ 66 \\ \hline \end{array}$

17. $\begin{array}{r} 1000 \\ \times\ 74 \\ \hline \end{array}$

18. $\begin{array}{r} 100 \\ \times\ 875 \\ \hline \end{array}$

Just the Facts! Workbook

Lesson 9: Multiplying by 1 and 11

Independent Work - Page 2

Do the problems below.

19.	20.	21.	22.
11 × 5	11 × 7	11 × 8	11 × 6

23.	24.	25.	26.
123 × 1	200 × 1	76 × 1	983 × 1

27. 11 x 4 = _____   30. 64 x 1 = _____

28. 5 x 11 = _____   31. 45 x 1 = _____

29. 3 x 11 = _____   32. 24 x 1 = _____

---

33. Emily has <u>four</u> bags of soccer balls. Each bag has <u>eleven</u> balls. How many soccer balls does Emily have?

____ x ____ = ____

Emily has

_____ soccer balls.

Lesson 10: Multiplying by 4

To multiply by four, we use the "**double double**" method.

For example: 3 x 4 = ☐

First, **double** the 3 to get 6, then **double** the 6 to get 12.

```
  3        6
+ 3      + 6         3 x 4 = 12
 ─        ──
  6       12
```

When we use the double double method, we often have to add two digit numbers. All you have to do is add the two columns separately.
Below, for 14+14 we add the 4+4 and the 1+1 to get 28.
* Note that regrouping (or carrying) is not covered in this book.

For example: 7 x 4 = ☐

```
  7       14
+ 7     + 14          7 x 4 = 28
 ──      ──
 14      28
```

(10.1)

4 x 4 = ☐

```
  4       8
+ 4     + 8
 ─       ─
  8
```

(10.2)

4 x 6 = ☐

```
  6       12
+ 6     + 12
 ─       ──
 12
```

Just the Facts! Workbook

Lesson 10: Multiplying by 4

Sometimes it's easier to memorize a multiplication fact with a simple story. For example, 4 x 8 maybe difficult to use the double double method since 16 + 16 may not be easy to do in your head.

4 x 8 = | 32 |

```
   8        16
 + 8      + 16
 ---      ----
  16        32
```

Instead, you can use the story below to remember this math fact.

Four ate (8) thirty-two.  There's 32, in 4's stomach.

Solve the four times equations below.

(10.3)
 	
   4
x 4
___
	(10.4)
	
	   4
	x 8
	___

(10.5)
 	
   3
x 4
___
	(10.6)
	
	   4
	x 6
	___

(10.7)
 	
   7
x 4
___
	(10.8)
	
	   4
	x 5
	___

Just the Facts! Workbook

Lesson 10: Multiplying by 4

Review: use your hands for the nine times problems below.

(10.9) 9 x 3 = _____   (10.13) 7 x 9 = _____

(10.10) 9 x 2 = _____   (10.14) 4 x 9 = _____

(10.11) 9 x 6 = _____   (10.15) 9 x 5 = _____

(10.12) 8 x 9 = _____   (10.16) 9 x 9 = _____

Review: count the zeros to solve the equations below.

(10.17) 50 x 100 = _____   (10.19) 710 x 100 = _____

(10.18) 10 x 60 = _____   (10.20) 100 x 600 = _____

Use the **Commutative Property** to rewrite the equations and solve.

(10.21) 5 x 4 = _____

(10.22) 2 + 8 = _____

(10.23) 7 x 4 = _____

(10.24) 9 + 3 = _____

Just the Facts! Workbook

Lesson 10: Multiplying by 4

Independent Work - Page 1

Solve the problems below.

1. 100 x 12 = _____
2. 100 x 100 = _____
3. 225 x 100 = _____

4. 64 x 1,000 = _____
5. 1000 x 23 = _____
6. 89 x 100 = _____

7. 9 x 9 = ___
8. 5 x 9 = ___
9. 9 x 3 = ___

10. 6 x 9 = ___
11. 4 x 9 = ___
12. 7 x 9 = ___

13. 5 x 5 = ___
14. 5 x 7 = ___
15. 6 x 5 = ___

16. 6 + 6 = ___
17. 5 + 5 = ___
18. 4 + 4 = ___

19. 3 + 3 = ___
20. 8 + 8 = ___
21. 7 + 7 = ___

Use the double double method to multiply by 4.

22.   4
    x 3

23.   7
    x 4

24.   6
    x 4

Just the Facts! Workbook

Lesson 10: Multiplying by 4

Independent Work - Page 2

25. Jack has <u>four</u> bags of beans. Each bag has <u>6</u> beans. How many beans does Jack have?

_____ x _____ = _____

Jack has _____ beans.

26. Jack has one bag of <u>ten</u> beans. He buys another bag of <u>five</u> beans. How many beans does Jack have in all?

_____ + _____ = _____

Jack has _____ beans.

27. Jack has <u>four</u> bags of beans. Each bag has <u>eight</u> beans? How many beans does Jack have?

_____ x _____ = _____

Jack has _____ beans.

Hint

Just the Facts! Workbook

Lesson 11: The Times Table Array

Below is the times table chart (or array), which shows all of the math facts that you must know.

Notice how each row or column **skip counts**.

If you want to multiply two numbers, find one number on the top row, and the other number on the left column, and use two fingers to go down the row and column, until your fingers meet. There's your answer. This is pretty cool, however you usually will not have this chart when doing problems.

	1	2	3	4	5	6	7	8	9	10	11
1	1	2	3	4	5	6	7	8	9	10	11
2	2	4	6	8	10	12	14	16	18	20	22
3	3	6	9	12	15	18	21	24	27	30	33
4	4	8	12	16	20	24	28	32	36	40	44
5	5	10	15	20	25	30	35	40	45	50	55
6	6	12	18	24	30	36	42	48	54	60	66
7	7	14	21	28	35	42	49	56	63	70	77
8	8	16	24	32	40	48	56	64	72	80	88
9	9	18	27	36	45	54	63	72	81	90	99
10	10	20	30	40	50	60	70	80	90	100	110
11	11	22	33	44	55	66	77	88	99	110	121

Just the Facts! Workbook

Lesson 11: The Times Table Array

The pink boxes, or lightly shaded boxes (if your book is not color) are the math facts you have learned.
You can see that you already know most of them!

	1	2	3	4	5	6	7	8	9	10	11
1	1	2	3	4	5	6	7	8	9	10	11
2	2	4	6	8	10	12	14	16	18	20	22
3	3	6	9	12	15	18	21	24	27	30	33
4	4	8	12	16	20	24	28	32	36	40	44
5	5	10	15	20	25	30	35	40	45	50	55
6	6	12	18	24	30	36	42	48	54	60	66
7	7	14	21	28	35	42	49	56	63	70	77
8	8	16	24	32	40	48	56	64	72	80	88
9	9	18	27	36	45	54	63	72	81	90	99
10	10	20	30	40	50	60	70	80	90	100	110
11	11	22	33	44	55	66	77	88	99	110	121

Use the chart above to find the answers, then circle how you would find the answers if you didn't have the chart.

(11.1) 7 x 4 = ☐   skip count by 7 | double double | use hands

(11.2) 5 x 6 = ☐   skip count by 5 | skip count by 6 | use hands

(11.3) 9 x 6 = ☐   skip count by 6 | skip count by 9 | use hands

Just the Facts! Workbook

Lesson 11: The Times Table Array

Review: solve the equations below.

(11.4) 9 x 9 = ____   (11.5) 11 x 10 = ____   (11.6) 3 x 3 = ____

(11.7) 5 x 5 = ____   (11.8) 4 x 8 = ____   (11.9) 76 x 1 = ____

Read the word problems below, and underline key information. Then write an equation, look for words such as "times" and "in all", and then solve.

(11.10) There are 52 weeks in a year. How many weeks are in 10 years?

(11.11) Jack has nine times as many pencils as Mary. Mary has 4 pencils. How many pencils does Jack have?

(11.12) Tim makes $6.00 an hour, and works 4 hours. How much did Tim earn?

(11.13) There are twenty-five students in a class. Each student has ten fingers. How many fingers do all the students have in all?

Just the Facts! Workbook

Lesson 11: The Times Table Array

Independent Work - Page 1

Without using the times table chart, solve the equations below.

1.
$$\begin{array}{r} 9 \\ \times 6 \\ \hline \end{array}$$

2.
$$\begin{array}{r} 9 \\ \times 9 \\ \hline \end{array}$$

3.
$$\begin{array}{r} 6 \\ \times 5 \\ \hline \end{array}$$

4.
$$\begin{array}{r} 4 \\ \times 3 \\ \hline \end{array}$$

5.
$$\begin{array}{r} 3 \\ \times 5 \\ \hline \end{array}$$

6.
$$\begin{array}{r} 4 \\ \times 5 \\ \hline \end{array}$$

7.
$$\begin{array}{r} 2 \\ \times 4 \\ \hline \end{array}$$

8.
$$\begin{array}{r} 4 \\ \times 9 \\ \hline \end{array}$$

9.
$$\begin{array}{r} 9 \\ \times 2 \\ \hline \end{array}$$

10.
$$\begin{array}{r} 2 \\ \times 3 \\ \hline \end{array}$$

11.
$$\begin{array}{r} 3 \\ \times 3 \\ \hline \end{array}$$

12.
$$\begin{array}{r} 7 \\ \times 5 \\ \hline \end{array}$$

13.
$$\begin{array}{r} 5 \\ \times 8 \\ \hline \end{array}$$

14.
$$\begin{array}{r} 9 \\ \times 4 \\ \hline \end{array}$$

15.
$$\begin{array}{r} 11 \\ \times 6 \\ \hline \end{array}$$

16.
$$\begin{array}{r} 100 \\ \times 54 \\ \hline \end{array}$$

17.
$$\begin{array}{r} 1000 \\ \times 48 \\ \hline \end{array}$$

18.
$$\begin{array}{r} 100 \\ \times 789 \\ \hline \end{array}$$

Lesson 11: The Times Table Array

Independent Work - Page 2

19. Fill in the missing numbers on the chart below.

	1	2	3	4	5	6	7	8	9	10	11
1	1	2	3		5	6	7		9	10	11
2	2	4	6	8	10	12		16	18		22
3	3	6				21	24			30	33
4		8		16	20		28			40	44
5	5	10	15	20		30	35	40		50	
6	6	12			30	36	42	48	54		66
7	7	14	21	28	35	42	49	56	63	70	77
8		16	24		40	48	56	64	72	80	
9	9	18			45	54	63	72		90	99
10	10	20	30	40	50		70	80	90	100	110
11	11		33	44		66	77		99	110	121

Just the Facts! Workbook

Lesson 12: Multiplication Double Facts

Below is the times table chart, where the green boxes (or, if your book is not color, the darker shaded boxes going diagonally) are the **multiplication <u>double</u> facts**, such as 2 times 2.

Even though you can skip count, or use other tricks (such as the nine times hand trick), you should try your best to commit the multiplication double facts to memory.

	1	2	3	4	5	6	7	8	9	10	11
1	1	2	3	4	5	6	7	8	9	10	11
2	2	4	6	8	10	12	14	16	18	20	22
3	3	6	9	12	15	18	21	24	27	30	33
4	4	8	12	16	20	24	28	32	36	40	44
5	5	10	15	20	25	30	35	40	45	50	55
6	6	12	18	24	30	36	42	48	54	60	66
7	7	14	21	28	35	42	49	56	63	70	77
8	8	16	24	32	40	48	56	64	72	80	88
9	9	18	27	36	45	54	63	72	81	90	99
10	10	20	30	40	50	60	70	80	90	100	110
11	11	22	33	44	55	66	77	88	99	110	121

Use the chart to fill in the answers.

(12.1) 1 x 1 = _____

(12.2) 2 x 2 = _____

(12.3) 3 x 3 = _____

(12.4) 4 x 4 = _____

(12.5) 5 x 5 = _____

(12.6) 6 x 6 = _____

(12.7) 7 x 7 = _____

(12.8) 8 x 8 = _____

(12.9) 9 x 9 = _____

(12.10) 10 x 10 = _____

Just the Facts! Workbook

Lesson 12: Multiplication Double Facts

Now we're going to concentrate on the multiplication double facts that are more difficult.

4 x 4 = 16   6 x 6 = 36   7 x 7 = 49   8 x 8 = 64

---

4 x 4 = 16

For a four times a number you use the **double double** method.

4 + 4 = 8

8 + 8 = **16**

If you like the simple story method, to arrive at an answer more quickly, you can use the story below.

The 4 twins celebrated their Sweet 16 Birthday!

Lesson 12: Multiplication Double Facts

6 x 6 = 36

Here, we can use the 5 times facts to help find the answer, since it's easy to skip count by 5.

6 x 5 = 5 + 5 + 5 + 5 + 5 + 5 = 30
(six groups of five)

using the **commutative property**, we know that we can rewrite this as

5 x 6 = 6 + 6 + 6 + 6 + 6 = 30
(five groups of six)

We need to add one more six to get the six groups of six.

6 x 6 = 6 + 6 + 6 + 6 + 6 + 6 = 36
⎣_____⎦
30

30 + 6 = **36**

If you like the simple story method, to arrive at an answer more quickly, you can use the story below.

Dad 6 and his son, Little 6, go fishing and catch 36 fish!

Just the Facts! Workbook

Lesson 12: Multiplication Double Facts

7 x 7 = 49

If you can't remember this math fact, you can use the story below to help you.

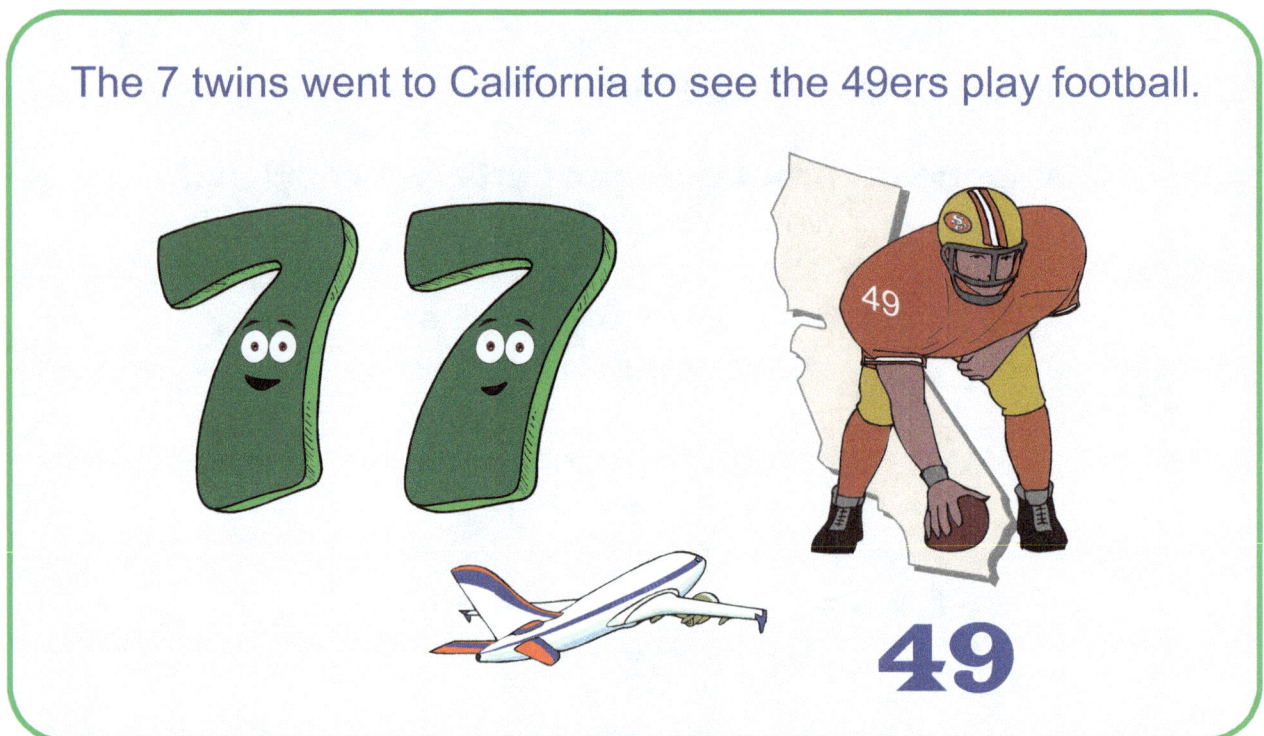

The 7 twins went to California to see the 49ers play football.

Solve the problems below.

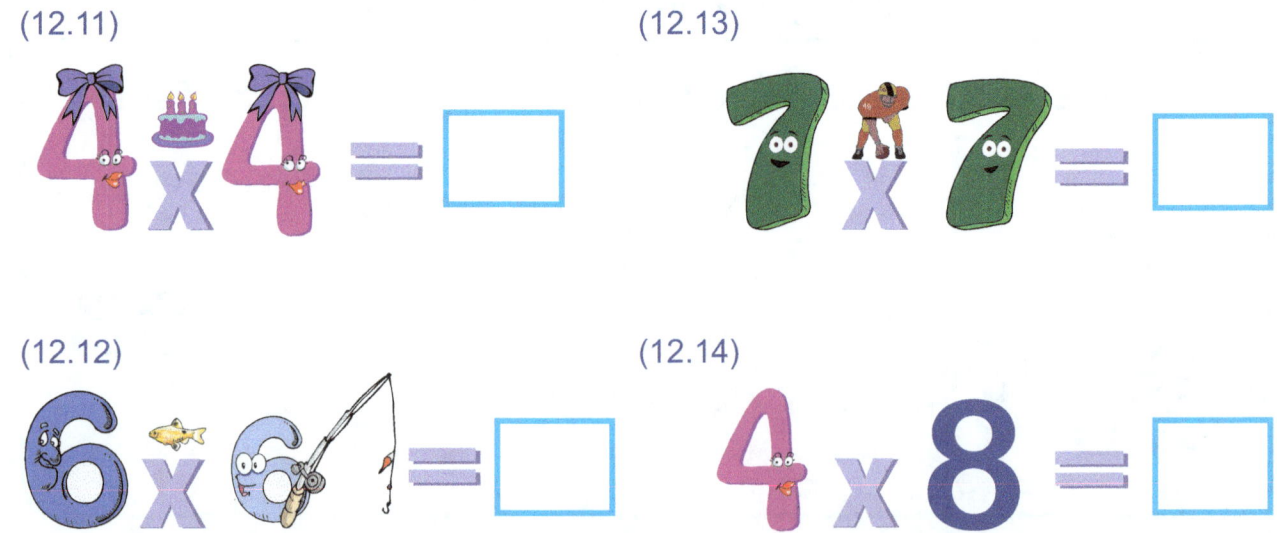

(12.11) 4 x 4 = ☐

(12.13) 7 x 7 = ☐

(12.12) 6 x 6 = ☐

(12.14) 4 x 8 = ☐

Just the Facts! Workbook

Lesson 12: Multiplication Double Facts

8 x 8 = 64

If you can't remember this math fact, you can use the story below.

8 and 8 went to the store, to buy Nintendo 64.

Solve the problems below.

(12.15)

(12.17)

(12.16)

(12.18)

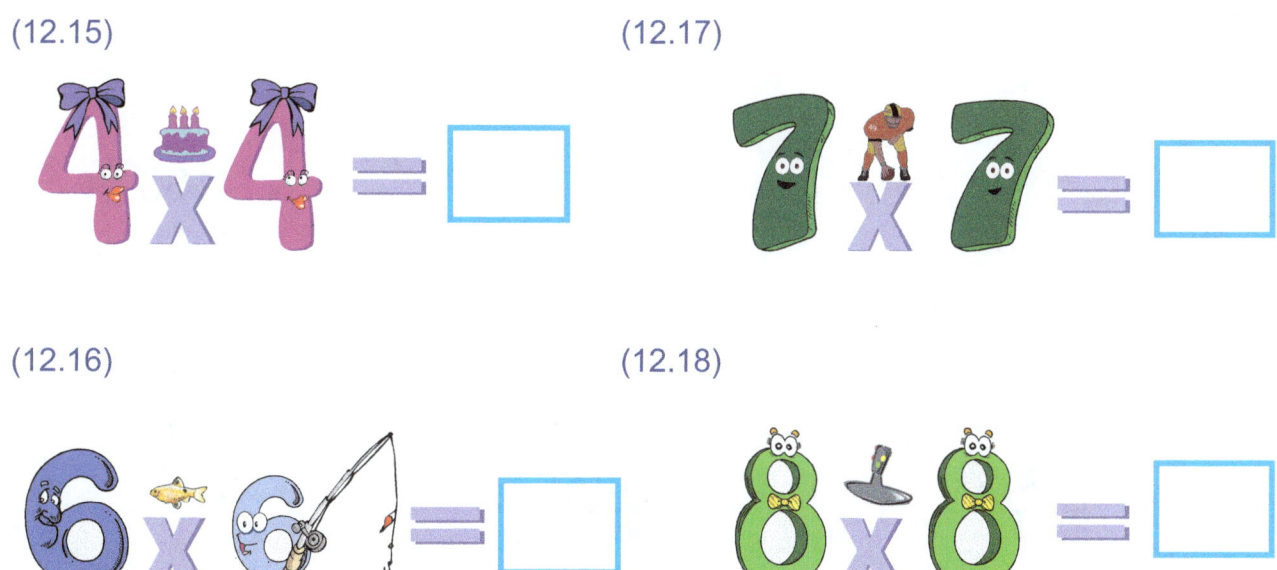

Just the Facts! Workbook

Lesson 12: Multiplication Double Facts

Independent Work - Page 1

Without using the times table chart, solve the equations below.

1.
```
   2
 x 2
```

2.
```
   2
 x 7
```

3.
```
   2
 x 8
```

4.
```
   2
 x 6
```

5.
```
   2
 x 3
```

6.
```
   2
 x 4
```

7.
```
   3
 x 3
```

8.
```
   3
 x 4
```

9.
```
   3
 x 9
```

10.
```
   4
 x 3
```

11.
```
   4
 x 5
```

12.
```
   4
 x 9
```

13.
```
    10
  x 8
```

14.
```
    10
  x 21
```

15.
```
   100
  x 10
```

16.
```
   100
 x 643
```

17.
```
   100
 x 490
```

18.
```
   100
 x 100
```

Just the Facts! Workbook

Lesson 12: Multiplication Double Facts

Independent Work - Page 2

Fill in the missing numbers for the math story facts below.

19.

20.

21.

22.

23.

24. There are seven days in a week. How many days are in seven weeks?

___ x ___ = ___

There are ___ days in 7 weeks.

25. Rewrite the equation below using multiplication, and solve.

4 + 4 + 4 + 4

___ x ___ = ___

26. Rewrite the equation below using multiplication, and solve.

3 + 3 + 3 + 3

___ x ___ = ___

Lesson 13: Multiplying by 3 Revisited

In a earlier lessons, you were taught to skip count by 3 when multiplying by 3. This can be difficult when multiplying 3 by larger numbers.
In this lesson, we will look at other ways to do the equations below.

3 x 6 = 18   3 x 7 = 21   3 x 8 = 24

3 x 6 = 18

To solve this equation, you can skip count by 3.

3, 6, 9, 12, 15, 18

or you can use the simple story method.

3 and 6 took bus 18 to school.

Just the Facts! Workbook

Lesson 13: Multiplying by 3 Revisited

3 x 7 = 21

For this equation, you can skip count by 3,

3, 6, 9, 12, 15, 18, 21

or double the 7, then add a third 7.

7 + 7 = 14,   14 + 7 = 21

The above can take some time,
so it's best to have this math fact memorized.

To help, you can use the simple story below.

7 had 3 sips of sparkling cider on his 21st birthday!

Lesson 13: Multiplying by 3 Revisited

$$3 \times 8 = 24$$

For this equation, you could skip count by 3,

$$3, 6, 9, 12, 15, 18, 21, 24$$

or double the 8, and then add a third 8.

$$8 + 8 = 16, \quad 16 + 8 = 24$$

Again, the above can take some time, so it's best to have this math fact memorized. To help, you can use the simple story below.

For this story, you need to know that there are 24 hours in 1 day.

Snowman 8 always eats three snow cones in 24 hours (one day).

Just the Facts! Workbook

Lesson 13: Multiplying by 3 Revisited

Fill in the missing numbers for the equations below.

(13.1)

(13.2)

(13.3)

(13.4)

(13.5)

(13.6)

(13.7)

(13.8) 4 x ☐ = 32

Review: solve the equations below.

(13.9) 1 x 1 = _____

(13.10) 2 x 2 = _____

(13.11) 3 x 3 = _____

(13.12) 4 x 4 = _____

(13.13) 5 x 5 = _____

(13.14) 6 x 6 = _____

(13.15) 7 x 7 = _____

(13.16) 8 x 8 = _____

(13.17) 9 x 9 = _____

(13.18) 10 x 10 = _____

Just the Facts! Workbook

Lesson 13: Multiplying by 3 Revisited

Independent Work - Page 1

1. 4 x 4 = ☐
2. 6 x 6 = ☐
3. 7 x 7 = ☐
4. 8 x 8 = ☐
5. 3 x 6 = ☐
6. 7 x 3 = ☐
7. 8 x 3 = ☐
8. 4 x 8 = ☐

9. There are seven days in a week. How many days are in three weeks?

10. Rewrite the equation below using multiplication, and solve:

    8 + 8 + 8 + 8

Just the Facts! Workbook

Lesson 13: Multiplying by 3 Revisited

Independent Work - Page 2

11. Fill in the missing numbers on the chart below.

	1	2	3	4	5	6	7	8	9	10	11
1		2	3	4	5	6	7	8	9	10	11
2	2		6	8	10	12	14	16	18	20	22
3	3	6		12	15				27	30	33
4	4	8	12		20	24	28		36	40	44
5	5	10	15	20		30	35	40	45	50	55
6	6	12		24	30		42	48	54	60	66
7	7	14		28	35	42		56	63	70	77
8	8	16			40	48	56		72	80	88
9	9	18	27	36	45	54	63	72		90	99
10	10	20	30	40	50	60	70	80	90		110
11	11	22	33	44	55	66	77	88	99	110	121

Lesson 14

7 x 6, 7 x 8, and 8 x 6

You now know most of the math facts in the chart below. Now let's take a look at the times table chart to see what's left.

	1	2	3	4	5	6	7	8	9	10	11
1	1	2	3	4	5	6	7	8	9	10	11
2	2	4	6	8	10	12	14	16	18	20	22
3	3	6	9	12	15	18	21	24	27	30	33
4	4	8	12	16	20	24	28	32	36	40	44
5	5	10	15	20	25	30	35	40	45	50	55
6	6	12	18	24	30	36	42	48	54	60	66
7	7	14	21	28	35	42	49	56	63	70	77
8	8	16	24	32	40	48	56	64	72	80	88
9	9	18	27	36	45	54	63	72	81	90	99
10	10	20	30	40	50	60	70	80	90	100	110
11	11	22	33	44	55	66	77	88	99	110	121

The math facts below are all that's left.

7 x 6 and 6 x 7   8 x 6 and 6 x 8   8 x 7 and 7 x 8
     = 42   = 48   = 56

Note that 11 x 11 will be covered later when we learn how to multiply double digits.

Lesson 14: 7 x 6, 7 x 8, and 8 x 6

> You can use other math facts to find an answer to a math fact that you forgot. You can always find an answer by using **multiplication double facts** (this is why it's good to have these memorized).

$$7 \times 6 = 42$$

We can use the 6 x 6 double fact and then add one more 6.

6 x 6 = 6 + 6 + 6 + 6 + 6 + 6 = 36
6 x 7 = 6 + 6 + 6 + 6 + 6 + 6 + 6

36

36 + 6 = 42

If you prefer math stores, use the rhyme below.

7 is green, 6 is blue, 7 times 6 is 42.

  42

If your book is not color, color the seven green and color the 6 blue.

Just the Facts! Workbook

Lesson 14: 7 x 6, 7 x 8, and 8 x 6

7 x 8 = 56

We can use the 7 x 7 double fact and then add one more 7.

7 x 7 = 7 + 7 + 7 + 7 + 7 + 7 + 7 = 49
7 x 8 = 7 + 7 + 7 + 7 + 7 + 7 + 7 + 7

49

49 + 7 = 56

Or you. can use the simple story method.

7 and 8 went for a drive, and got pulled over for going **one mile** over the speed limit.

Just the Facts! Workbook

Lesson 14: 7 x 6, 7 x 8, and 8 x 6

8 x 6 = 48

We can use the 6 x 6 double fact to find the answer, but then you have to add 6 **two** more times.

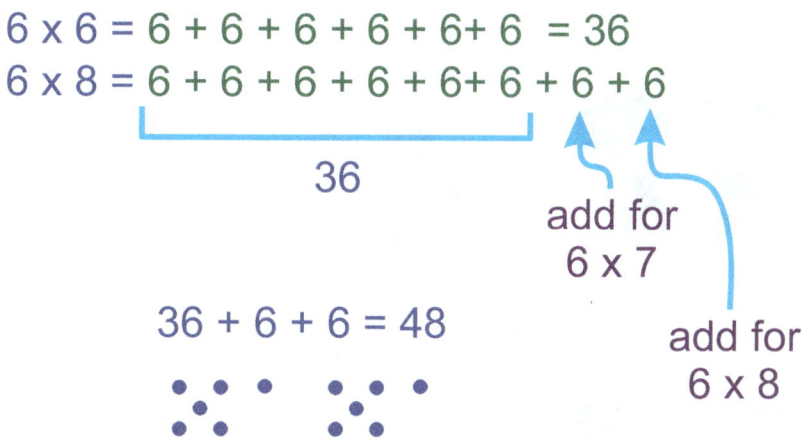

36 + 6 + 6 = 48

Or you can use the simple story method.

6 built snowman 8 and made him 48 inches tall.

Just the Facts! Workbook

Lesson 14: 7 x 6, 7 x 8, and 8 x 6

Now let's see if you can remember the stories. Look at the stories below and say the story out loud, and then write the multiplication equation with the answer that goes with the story. Note that the stories are not complete, so you will have to fill in the missing pieces.

(14.1) ___ x ___ = ___

(14.2) ___ x ___ = ___

(14.3) ___ x ___ = ___

(14.4) ___ x ___ = ___

Just the Facts! Workbook

Lesson 14: 7 x 6, 7 x 8, and 8 x 6

Continued from previous page.

(14.5) Hint: 7 is green and 6 is blue.

___ x ___ = ___

(14.6)

___ x ___ = ___

(14.7)

___ x ___ = ___

(14.8)

___ x ___ = ___

(14.9)

___ x ___ = ___

Just the Facts! Workbook

Lesson 14: 7 x 6, 7 x 8, and 8 x 6

Continued from previous page.

(14.10)    ___ x ___ = ___

(14.11)      ___ x ___ = ___

Review: solve the equations below.

(14.12) 5 x 1 = _____   (14.17) 5 x 6 = _____

(14.13) 5 x 2 = _____   (14.18) 5 x 7 = _____

(14.14) 5 x 3 = _____   (14.19) 5 x 8 = _____

(14.15) 5 x 4 = _____   (14.20) 5 x 9 = _____

(14.16) 5 x 5 = _____   (14.21) 5 x 10 = _____

(14.22) 9 x 9 = _____   (14.25) 2 x 6 = _____

(14.23) 9 x 8 = _____   (14.26) 2 x 7 = _____

(14.24) 9 x 7 = _____   (14.27) 2 x 8 = _____

Lesson 14: 7 x 6, 7 x 8, and 8 x 6

Independent Work - Page 1

Fill in the answers for the stories below.

Lesson 14: 7 x 6, 7 x 8, and 8 x 6

Independent Work - Page 2

Solve the equations below.

12. 4 x 8 = ____	13. 6 x 6 = ____
14. 7 x 8 = ____	15. 8 x 3 = ____
16. 3 x 6 = ____	17. 4 x 4 = ____
18. 8 x 8 = ____	19. 7 x 3 = ____
20. 7 x 6 = ____	21. 7 x 7 = ____
22. 6 x 8 = ____	

Solve the equations below.

23. 3   24. 4   25. 9   26. 5   27. 4
   x 3   x 3   x 3   x 5   x 5
   ___   ___   ___   ___   ___

28. Fill in the missing part of the story for 6 x 8.

Just the Facts! Workbook

Lesson 15: Multiplying by Zero

Multiplying by zero may be obvious when you think of it in terms of groups.

If you have zero groups of a number, you have zero.

0 group of **5** dots, or

0 x 5 = 0

Anything times zero will always be zero.

(15.1) 5 x 0 = _____

(15.4) 345 x 0 = _____

(15.2) 0 x 2 = _____

(15.5) 0 x 1,598 = _____

(15.3) 0 x 31 = _____

(15.6) 693 x 0 = _____

Review: for the pictures below, write the equations and solve.

(15.7)

____ groups of ____ beans

____ x ____ = ____

(15.8)

____ groups of ____ cents

____ x ____ = ____

(15.9)

____ groups of ____ fish

____ x ____ = ____

(15.10)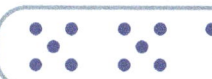

____ groups of ____ dots

____ x ____ = ____

Lesson 15: Multiplying by Zero

Review: fill in the answers for the stories below.

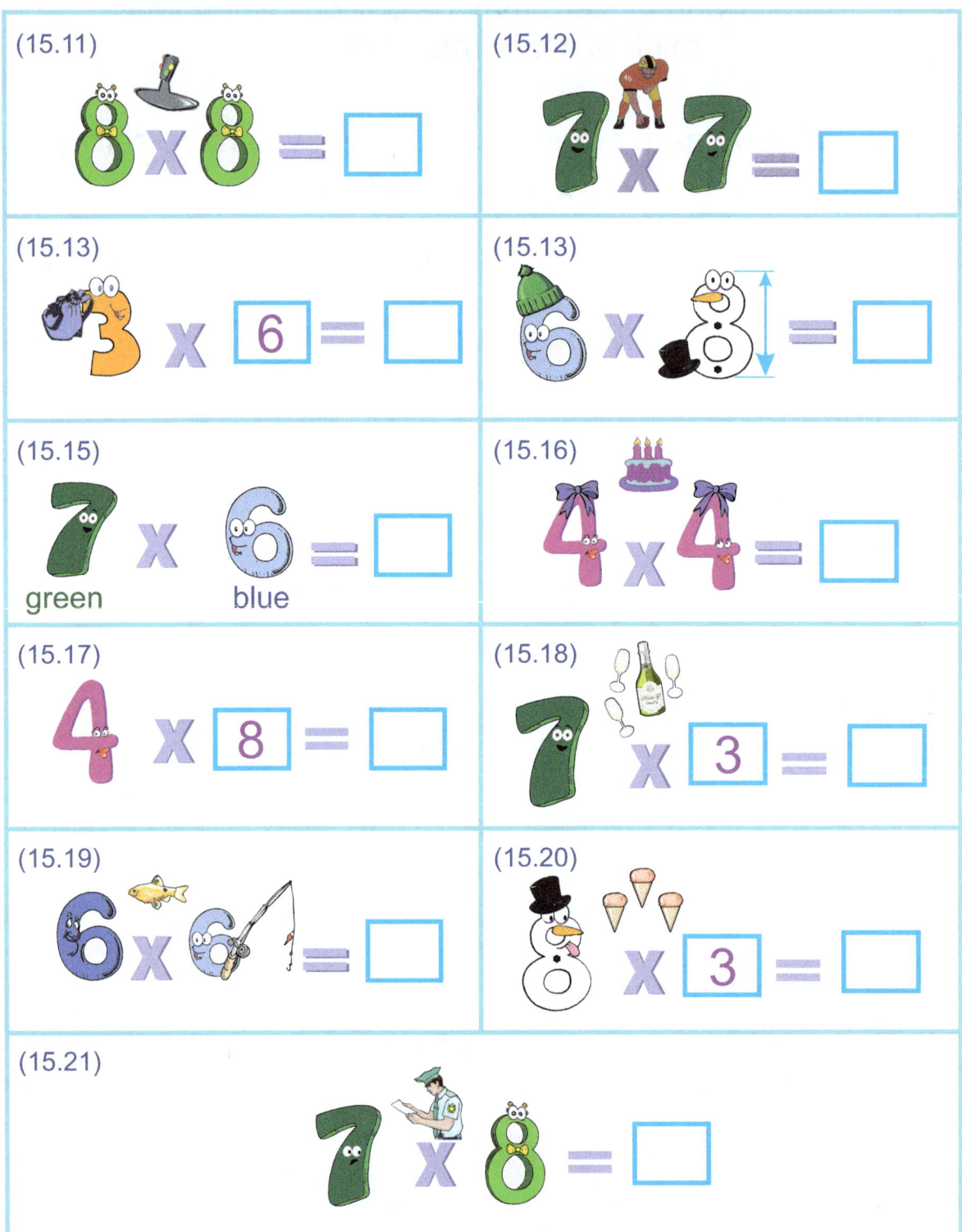

Lesson 15: Multiplying by Zero

Independent Work - Page 1

Solve the equations below.

1. $3 \times 0 =$ _____
2. $11 \times 1 =$ _____
3. $64 \times 0 =$ _____
4. $0 \times 1{,}000 =$ _____
5. $1 \times 52 =$ _____
6. $988 \times 0 =$ _____

7. 9 × 9	8. 5 × 5	9. 3 × 3	10. 2 × 2	11. 4 × 4
12. 3 × 2	13. 2 × 5	14. 2 × 3	15. 2 × 4	16. 9 × 2
17. 7 × 2	18. 2 × 8	19. 5 × 4	20. 8 × 5	21. 0 × 5
22. 9 × 6	23. 7 × 9	24. 9 × 0	25. 4 × 7	26. 4 × 5

Just the Facts! Workbook

Lesson 15: Multiplying by Zero

Independent Work - Page 2

27.  3 x 8 = ___

28.  7 x 7 = ___

29. 4 x 8 = ___

30. 8 x 8 = ___

31. 8 x 7 = ___

32. 6 x 6 = ___

33. green blue
7 x 6 = ___

34. 8 x 6 = ___

35.  7 x 3 = ___

36. 4 x 4 = ___

37. 3 x 6 = ___

Lesson 16: Factors and Products

All of the numbers in a multiplication equation have names. See the equation below.

$$\underline{5} \times \underline{6} = \underline{30}$$

These are called **factors**. This is called the **product**.

To help remember these words, you can think that the **factors** make the **product**, much like a **factory** (which has the word factor) makes **products**.

Circle the names for the underlined numbers.

(16.1) 5 x 3 = **15**   factor   product

(16.2) 8 x **8** = 64   factor   product

(16.3) **3** x 8 = 24   factor   product

(16.4) 6 x **8** = 48   factor   product

(16.5) 8 x 7 = **56**   factor   product

(16.6) 6 x 6 = **36**   factor   product

Just the Facts! Workbook

Lesson 16: Factors and Products

Below are other ways to write multiplication equations.

5 * 6 = 30   (5)(6) = 30

5 • 6 = 30   5(6) = 30

Is the underlined number a factor or a product?

(16.7) 3 * **6** = 18 factor product	(16.11) 6 * 6 = **36** factor product
(16.8) 8 • 6 = **48** factor product	(16.12) **9** • 6 = 54 factor product
(16.9) (**8**)(7) = 56 factor product	(16.13) (**5**)(5) = 25 factor product
(16.10) 7(**7**) = 49 factor product	(16.14) 3(8) = **24** factor product

Solve the equations below.

(16.15)   6 x 6 = ____

(16.18)   4(4) = ____

(16.21)   8 * 4 = ____

(16.16)   8 • 8 = ____

(16.19)   6 • 4 = ____

(16.22)   7 * 4 = ____

(16.17)   (7)(7) = ____

(16.20)   (7)(8) = ____

(16.23)   3(8) = ____

Just the Facts! Workbook

Lesson 16: Factors and Products

Independent Work - Page 1

Fill in the answers for the stories below.

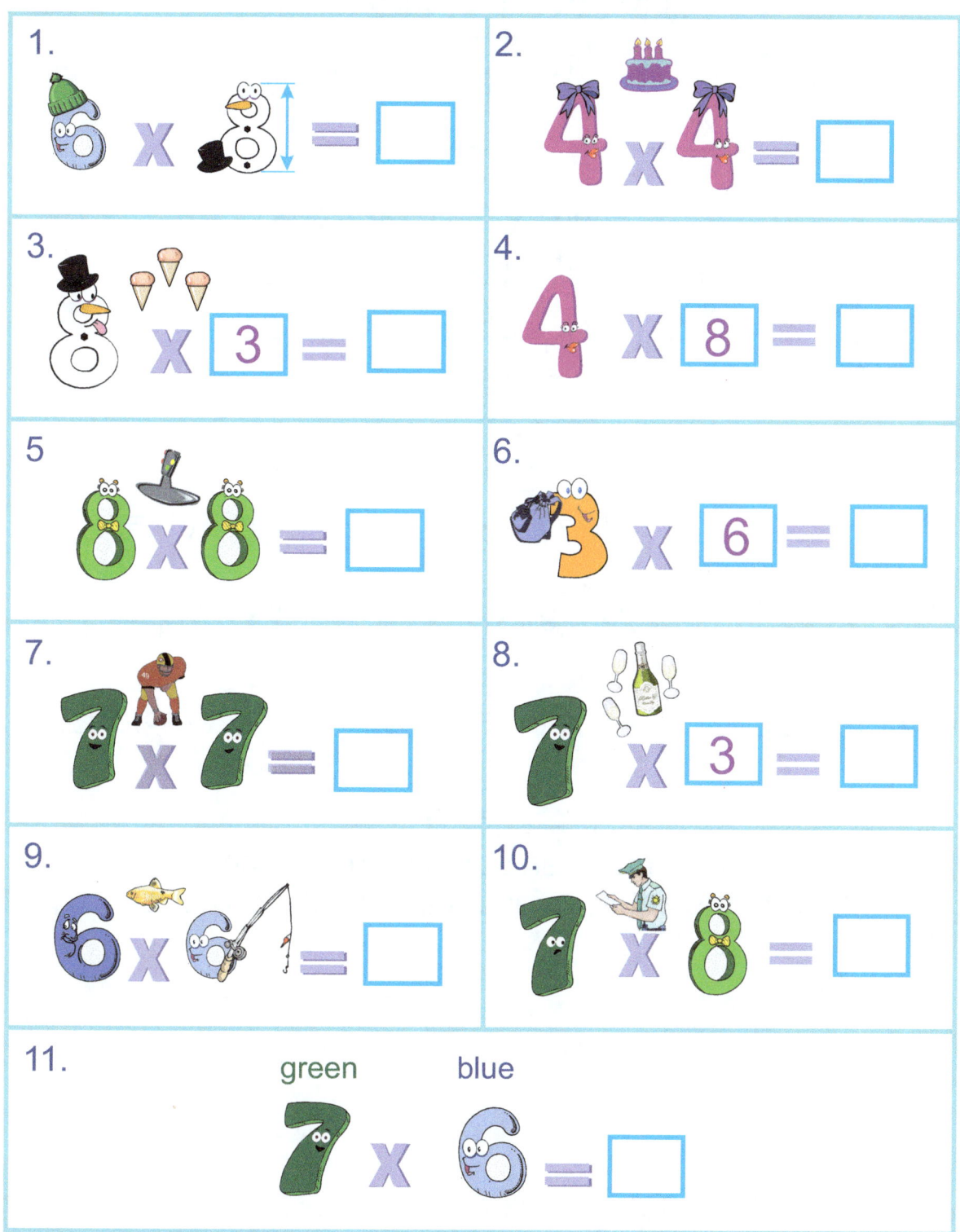

Lesson 16: Factors and Products

Independent Work - Page 2

Solve the equations below.

12. 11 × 3	13. 10 × 3	14. 100 × 4
15. 77 × 1	16. 99 × 0	17. 112 × 0

18. 5 * 3 = _____

19. 4(4) = _____

20. (9)(9) = _____

21. 7 * 8 = _____

22. (9)(3) = _____

23. 3(8) = _____

24. 7(4) = _____

25. 7 • 7 = _____

26. (9)(8) = _____

27. The factors in an equation are 4 and 6.
Write the equation and then find the product.

Just the Facts! Workbook

Lesson 17 — Commas in Large Numbers

Before we learn how to multiply large numbers, you need to know the place value of commas and numbers within a larger number.

981,123,456,789

The number above is:
nine hundred eighty one billion,
one hundred twenty-three million,
four hundred fifty-six thousand,
seven hundred eighty-nine.

Use the commas to help you put the number into words. Each comma has a name, depending on where it is.

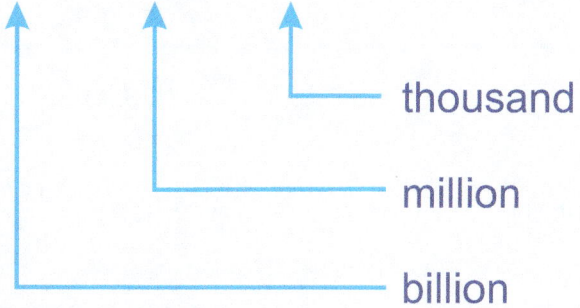

Notice how you read each group of three numbers before the comma.

nine hundred eight-one	981
one hundred twenty-three	123
four hundred fifty-six	456
seven hundred eighty-nine	789

Just the Facts! Workbook

Lesson 17: Commas in Large Numbers

Circle the comma name for the indicated comma.

(17.1) 152,561,122 ↑ (million comma)

| billion | million | thousand |

(17.2) 1,891,000 ↑ (million comma)

| billion | million | thousand |

(17.3) 45,012,213 ↑ (million comma)

| billion | million | thousand |

(17.4) 1,000,000 ↑ (million comma)

| billion | million | thousand |

(17.5) 43,000,000 ↑ (million comma)

| billion | million | thousand |

(17.6) 34,431,968,000 ↑ (billion comma)

| billion | million | thousand |

(17.7) 132,547,973 ↑ (million comma)

| billion | million | thousand |

Just the Facts! Workbook

Lesson 17: Commas in Large Numbers

How to Read a Number

981,123,456,789

To read a large number, break it up into segments of three digits of hundred/tens/ones.

hundred / ten / one

981,123,456,789

- 981 — nine hundred eighty-one
- 123 — one hundred twenty-three
- 456 — four hundred fifty-six
- 789 — seven hundred eighty-nine

It's the commas that tell you the next word for the word number.

981,123,456,789

- 981 billion — nine hundred eighty-one
- 123 million — one hundred twenty-three
- 456 thousand — four hundred fifty-six
- 789 — seven hundred eighty-nine

Read the number in words out loud.

Just the Facts! Workbook

Lesson 17: Commas in Large Numbers

Circle the number that matches the word.

(17.8) forty-five thousand one hundred two

    45,102   45,120   450,102

(17.9) one hundred sixty-five thousand, five hundred

    16,500   165,000   165,500

(17.10) one hundred million, one hundred twenty-three thousand, one

    100,123   100,123,001   100,123,100

(17.11) sixty-five billion

    65,000   65,000,000   65,000,000,000

Write the word number on the lines provided.
Note: Never use the word "**and**", and add a hyphen between tens and ones digits, such as: **twenty-five**.

(17.12)   8,700,120

_____

_____

(17.13)   41,302,000

_____

_____

Lesson 17: Commas in Large Numbers

Independent Work - Page 1

1. Draw lines to match the words to their commas.

   1 2 3 , 4 5 6 , 7 8 9 , 3 2 1

   million   thousand   billion

Is the number the same as the word number?

2.	65,102 sixty-five thousand, one hundred two
3.	1,600,153 one billion, six hundred thousand, one hundred fifty-three
4.	756,412,000 seven hundred fifty-six million, four hundred twelve thousand
5.	89,000 eighty-nine thousand
6.	51,000,001 fifty-one thousand, one

Lesson 17: Commas in Large Numbers

Independent Work - Page 2

Fill in the answers for the stories below.

7. 

8.  x

9.

10.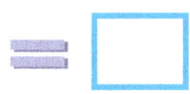

11.  x 3 =

12.  x

13.

14.  x 6 =

15.  x

16.  x 8 =

17.  x 3 =

Lesson 18: Place Value of Numbers

Not only do commas have names, but each digit inside a large number has a name, or, more importantly, a value.

Again, you can split the large number up into segments of three digits.

Each digit has a hundred, a ten, and a one value. The comma to the right of the segment will tell you the rest of the name.

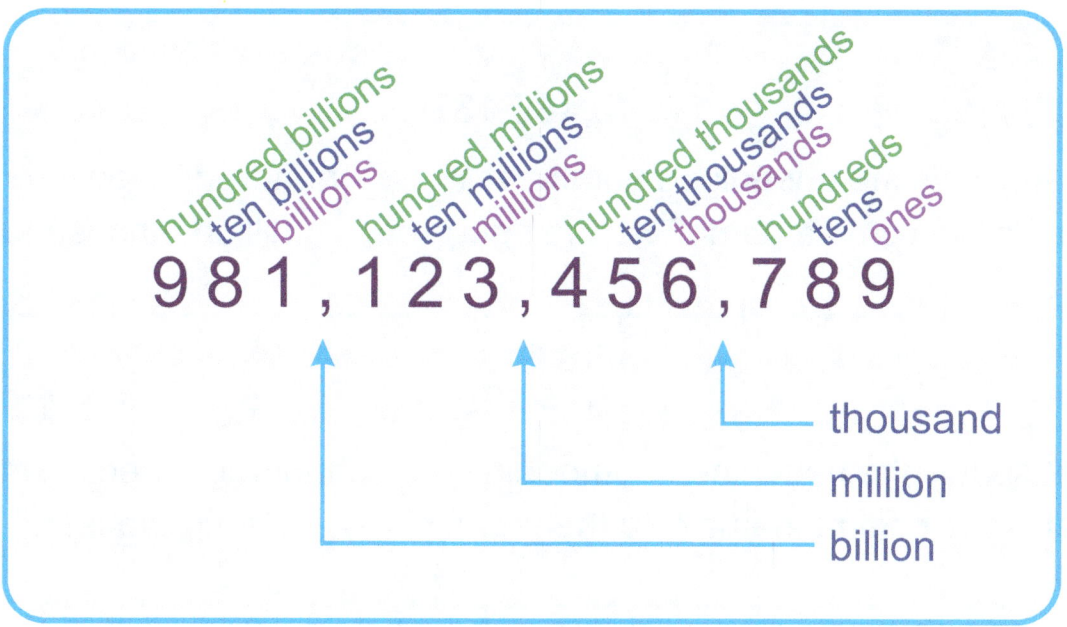

Example: **5<u>6</u>**

The underlined number is a **ten** number.
There is no comma to the right of the number, therefore, this number is simply in the **tens** spot.

Example: **981,<u>1</u>23,456,789**

The underlined number is a **hundred** number.
The **comma** to the right of that number is the **million comma**.
Therefore, the underlined number is in the **hundred millions** spot.

Just the Facts! Workbook

Lesson 18: Place Value of Numbers

Answer the questions, and then say (out loud) the name
of the spot that the underlined number is in.
The first one was done for you.

(18.1)

5<u>6</u>3,487

What spot is the underlined number in?   hundred / (ten) / one
What is the comma name? (to the right)   million / (thousand)

"The 6 is in the ten thousands spot."

(18.2)

<u>2</u>,486,137

What spot is the underlined number in?   hundred / ten / one
What is the comma name? (to the right)   million / thousand

(18.3)

8<u>7</u>4,125

What spot is the underlined number in?   hundred / ten / one
What is the comma name? (to the right)   million / thousand

(18.4)

<u>8</u>79,258,146

What spot is the underlined number in?   hundred / ten / one
What is the comma name? (to the right)   million / thousand

(18.5)

<u>8</u>,798

What spot is the underlined number in?   hundred / ten / one
What is the comma name? (to the right)   million / thousand

Lesson 18: Place Value of Numbers

Independent Work - Page 1

1. Draw lines to match the words to their commas.

$$123,456,789,321$$

million   thousand   billion

Circle the comma name for the indicated comma.

2.		459,000,188 ↑	
	billion	million	thousand
3.		89,877,400,221 ↑	
	billion	million	thousand
4.		8,000,000,000 ↑	
	billion	million	thousand
5.		578,788,000 ↑	
	billion	million	thousand

Just the Facts! Workbook

Lesson 18: Place Value of Numbers

Independent Work - Page 2

Draw lines to match the underlined number to the name of the spot that the number is in.

6. <u>1</u>,250   hundred thousands

7. 41,8<u>7</u>2   tens

8. <u>2</u>1,023   thousands

9. 34,<u>9</u>90   hundreds

10. <u>7</u>62,901   ten thousands

Solve the equations below.

11. 6 * 7 = _____   15. 6 x 6 = _____

12. (7)(8) = _____   16. 8 * 8 = _____

13. 9 * 8 = _____   17. (4)(4) = _____

14. 7 x 7 = _____   18. 5 x 5 = _____

Lesson 19: Multiplying Multiple Digits by Single Digits

To multiply larger numbers, you break it down into smaller equations.

Multiplying a 2 digit number by a 1 digit number.

$$\begin{array}{r} 25 \\ \times\ 3 \\ \hline \end{array}$$

$$\begin{array}{r} ^3\ \ \\ 26 \\ \times\ 5 \\ \hline 0 \end{array}$$

1) 5 x 6 = <u>3</u>0 (carry the three to the next column)

$$\begin{array}{r} 26 \\ \times\ 5 \\ \hline 130 \end{array}$$

2) 5 x 2 = 10 + <u>3</u> = 13

3) Answer: 130

Multiplying a 3 digit number by a 1 digit number.

$$\begin{array}{r} 250 \\ \times\ 3 \\ \hline \end{array}$$

$$\begin{array}{r} 250 \\ \times\ 3 \\ \hline 0 \end{array}$$

1) 0 x 3 = 0

$$\begin{array}{r} ^1\ \ \ \\ 250 \\ \times\ 3 \\ \hline 50 \end{array}$$

2) 5 x 3 = <u>15</u> (carry the one to the next column)

$$\begin{array}{r} 250 \\ \times\ 3 \\ \hline 750 \end{array}$$

3) 2 x 3 = 6 + <u>1</u> = 7

4) Answer: 750

Just the Facts! Workbook

Lesson 19: Multiplying Multiple Digits by Single Digits

Solve the problems below.

(19.1)

$$\begin{array}{r} 45 \\ \times\ 4 \\ \hline \end{array}$$

$$\begin{array}{r} 4\overline{5} \\ \times\ 4 \\ \hline \square \end{array}$$
5 x 4 = _____, do you need to carry a number?

$$\begin{array}{r} \overset{2}{4}5 \\ \times\ 4 \\ \hline \square\square 0 \end{array}$$
4 x 4 = _____, do you need to add a number?

(19.2)

$$\begin{array}{r} 536 \\ \times\ 2 \\ \hline \end{array}$$

$$\begin{array}{r} 536 \\ \times\ 2 \\ \hline \square \end{array}$$
6 x 2 = _____, do you need to carry a number?

3 x 2 = _____, do you need to add a number?

5 x 2 = _____, do you need to add a number?

Just the Facts! Workbook

Lesson 19: Multiplying Multiple Digits by Single Digits

Do the problems below:

(19.3)　22　× 3

(19.4)　32　× 2

(19.5)　12　× 3

Do the problems below, note that you will have to carry.

(19.6)　24　× 3

(19.7)　14　× 4

(19.8)　25　× 3

(19.9)　53　× 4

(19.10)　82　× 9

(19.11)　75　× 4

(19.12)　231　× 4

Just the Facts! Workbook

Lesson 19: Multiplying Multiple Digits by Single Digits

Independent Work - Page 1

1. 13
   x 2

2. 22
   x 2

3. 31
   x 3

Do the problems below, note that you will have to carry.

4. 24
   x 4

5. 13
   x 5

6. 35
   x 2

7. 62
   x 8

8. 55
   x 9

9. 43
   x 4

10. 222
    x 4

Just the Facts! Workbook

Lesson 19: Multiplying Multiple Digits by Single Digits

Independent Work - Page 2

Circle the comma name for the indicated comma.

11.	100,000,000 ↑		
	billion	million	thousand
12.	7,000,000,000 ↑		
	billion	million	thousand
13.	1,000,000,000 ↑		
	billion	million	thousand

Solve the problems below.

14. 88 × 0

15. 256 × 0

16. 510 × 1

17. 125 × 1

18. 22 × 10

19. 100 × 21

20. 100 × 20

Just the Facts! Workbook

Lesson 20: More on Multiplying Multiple Digits

To multiply larger numbers, you do the same as done in the last lesson: break it down into smaller equations.

Multiplying a 4 digit number by a 1 digit number.

```
    2,531
    x  3
```

```
   2,531           2,531         ¹            ¹
   x  3            x  3        2,531        2,531
   -----           -----       x   3        x   3
       3              93       -----        -----
                                 593         7593
```

You can use this method when multiplying a single digit by another number that has any number of digits.

Solve the problems below.

(20.1)   2,234
         x   2
         ▢▢▢▢

(20.2)   1,424
         x   3
         ▢▢▢▢

(20.3)   5,064
         x   4
         ▢▢▢▢▢

(20.4)   9,170
         x   3
         ▢▢▢▢▢

Lesson 20: More on Multiplying Multiple Digits

Solve the equations below.

(20.5) 8 * 3 = _____   (20.8) 3 * 3 = _____   (20.11) 8(4) = _____

(20.6) 3(4) = _____   (20.9) (4)(4) = _____   (20.12) 6 * 6 = _____

(20.7) (8)(8) = _____   (20.10) 7(8) = _____   (20.13) (7)(7) = _____

Is the underlined number a **factor** or a **product**?

(20.14) 4 * 6 = **24** factor product	(20.15) 8 * 8 = **64** factor product
(20.16) **7** · 6 = 42 factor product	(20.17) **6** · 8 = 48 factor product
(20.18) (4)(**8**) = 32 factor product	(20.19) (7)(3) = **21** factor product
(20.20) 9(7) = **63** factor product	(20.21) 7(**4**) = 28 factor product

Just the Facts! Workbook

Lesson 20: More on Multiplying Multiple Digits

Independent Work - Page 1

Solve the equations below.

1. $8 * 6 =$ ____
2. $4 * 7 =$ ____
3. $8 * 4 =$ ____

4. $4 \cdot 8 =$ ____
5. $8 \cdot 7 =$ ____
6. $4 \cdot 7 =$ ____

7. $(4)(4) =$ ____
8. $(3)(7) =$ ____
9. $(7)(7) =$ ____

10. $7(6) =$ ____
11. $8(3) =$ ____
12. $7(8) =$ ____

Solve the equations below.

13.

14.

15.

16.

Just the Facts! Workbook

Lesson 20: More on Multiplying Multiple Digits

Independent Work - Page 2

17. There are seven days in a week.
How many days are in eight weeks?

18. Rewrite the equation below using multiplication, and solve.

    6 + 6 + 6 + 6 =

19. There are twelve eggs in carton.
How many eggs are in 3 cartons?

20. Each table comes with six chairs.
How many chairs do we have if there are seven tables?

Just the Facts! Workbook

Lesson 21: Multiplying Multiple Digits by 2 Digit Numbers

To multiply multiple digit numbers by two digit numbers, you start out following the steps that you used before for single digits.

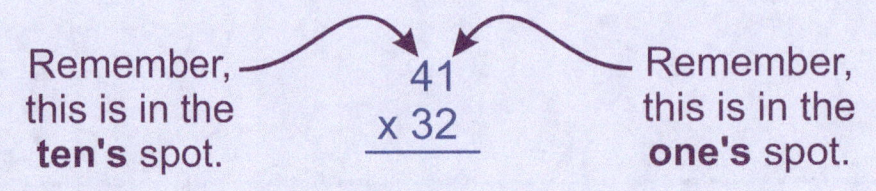

Remember, this is in the **ten's** spot. — 41 × 32 — Remember, this is in the **one's** spot.

Step 1: Multiply the number in the one's column on the bottom row with all numbers on the top row.

$$\begin{array}{r} 41 \\ \times 32 \\ \hline 2 \end{array} \qquad \begin{array}{r} 41 \\ \times 32 \\ \hline 82 \end{array} \qquad \begin{array}{l} 2 \times 1 = 2 \\ 2 \times 4 = 8 \end{array}$$

Step 2: Next we will multiply the ten's number on the bottom row, but first we must put a zero (a place holder) in the one's column for the answer.

$$\begin{array}{r} 41 \\ \times 32 \\ \hline 82 \\ 0 \end{array}$$

0 ← Place holder zero.

Step 3: Now multiply the ten's number in the second row with all the numbers in the top row.

$$\begin{array}{r} 41 \\ \times 32 \\ \hline 82 \\ 30 \end{array} \quad 3 \times 1 = 3 \qquad \begin{array}{r} 41 \\ \times 32 \\ \hline 82 \\ 1230 \end{array} \quad 3 \times 4 = 12$$

Just the Facts! Workbook

Lesson 21: Multiplying Multiple Digits by 2 Digit Numbers

Step 4: Lastly, add the two rows together to get the final answer.

```
    41              41
  x 32            x 32
   ───             ───
    82              82
 +1230           +1230
                  ────
                  1312
```
Add these rows.

41 x 32 = 1,312

Solve the equations below.

(21.1)
```
    21
  x 14
  ────
  □ □
+ □ □ 0
  ─────
  □ □ □
```

(21.2)
```
    43
  x 21
  ────
  □ □
+ □ □ 0
  ─────
  □ □ □
```

(21.3)
```
    31
  x 14
  ────
  □ □
+ □ □ 0
  ─────
  □ □ □
```

Lesson 21: Multiplying Multiple Digits by 2 Digit Numbers

Sometimes you will have to <u>carry</u> a number into another column.

(21.4)

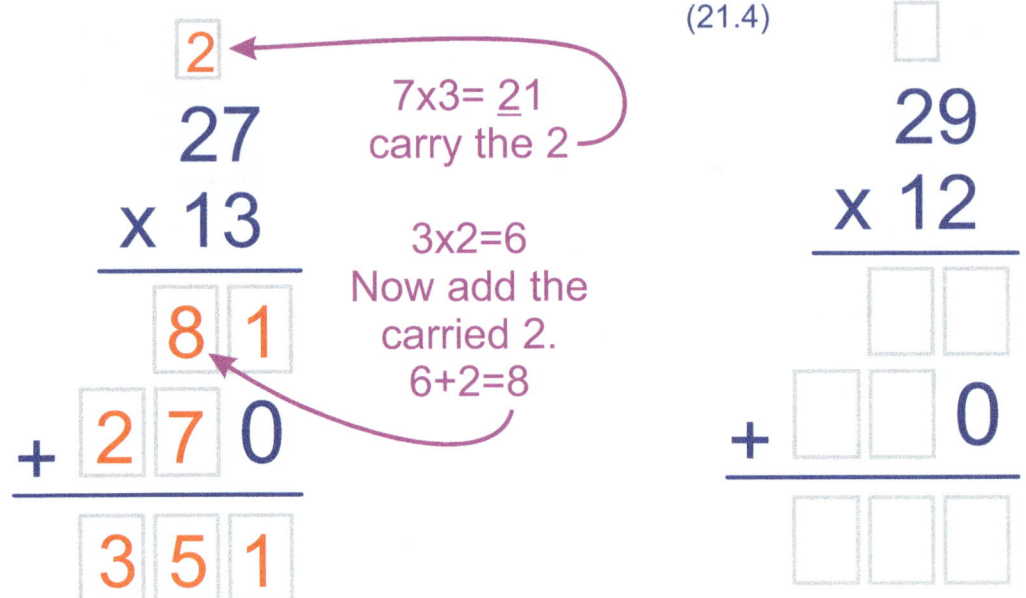

Above, we carried the ten's number for the one's number product. Sometimes you will have to carry for the ten's number, or both the one's *and* the ten's number product.

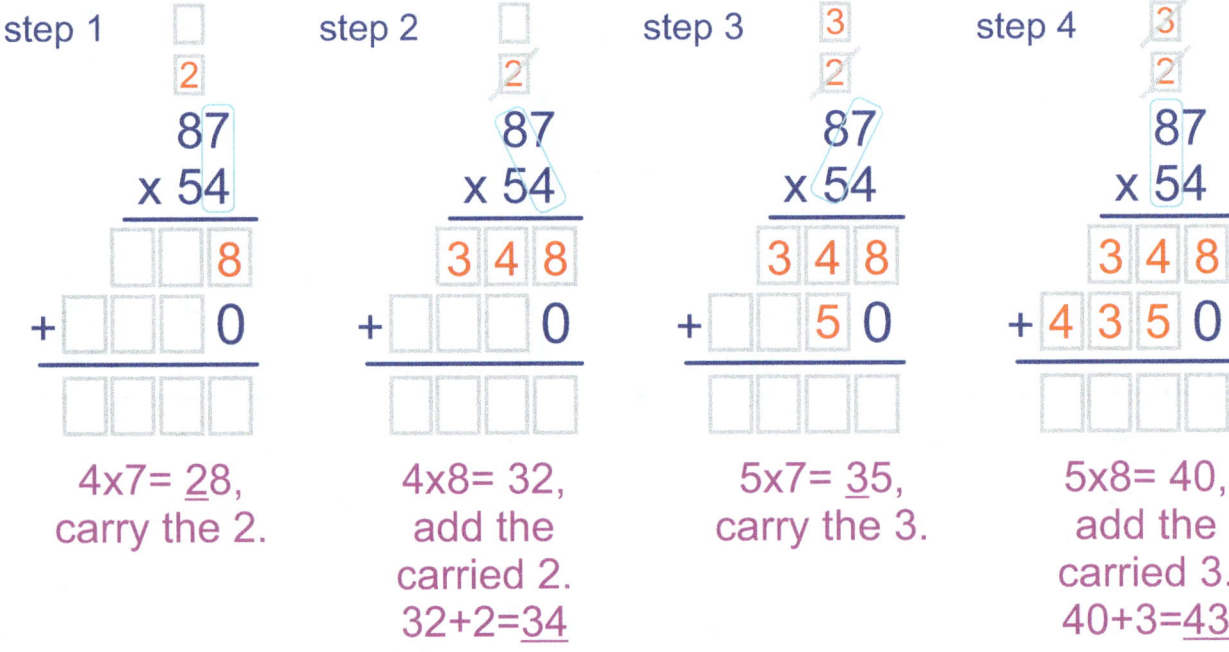

step 1	step 2	step 3	step 4
4x7= 2̲8, carry the 2.	4x8= 32, add the carried 2. 32+2=3̲4	5x7= 3̲5, carry the 3.	5x8= 40, add the carried 3. 40+3=4̲3

Next, add 348+4,350 to get the final answer.

Just the Facts! Workbook

Lesson 21: Multiplying Multiple Digits by 2 Digit Numbers

Solve the equations below.

(21.5)
$$53 \times 49$$

(21.6)
$$28 \times 36$$

(21.7)
$$82 \times 36$$

(21.8)
$$39 \times 94$$

Lesson 21: Multiplying Multiple Digits by 2 Digit Numbers

Independent Work - Page 1 of 1

Solve the equations below.

1. $\quad 33 \times 21$

2. $\quad 41 \times 32$

3. $\quad 53 \times 12$

4. $\quad 23 \times 35$

5. $\quad 87 \times 18$

Are the red numbers, in the equations, a factor or a product?

6. $\quad 62 \times 12 = 744$ (a) factor   (b) product

7. $\quad 45 \times 64 = 2{,}880$ (a) factor   (b) product

Just the Facts! Workbook

Lesson 22: Multiplying Multiple Digits by Three Digit Numbers

To multiply multiple digit numbers by three digit numbers, you do the same steps that you did when multiplying by two digits, however, you will now have a hundreds row to add.

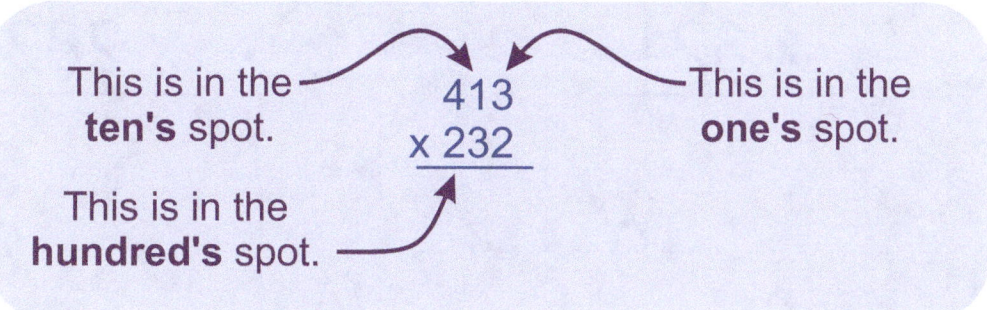

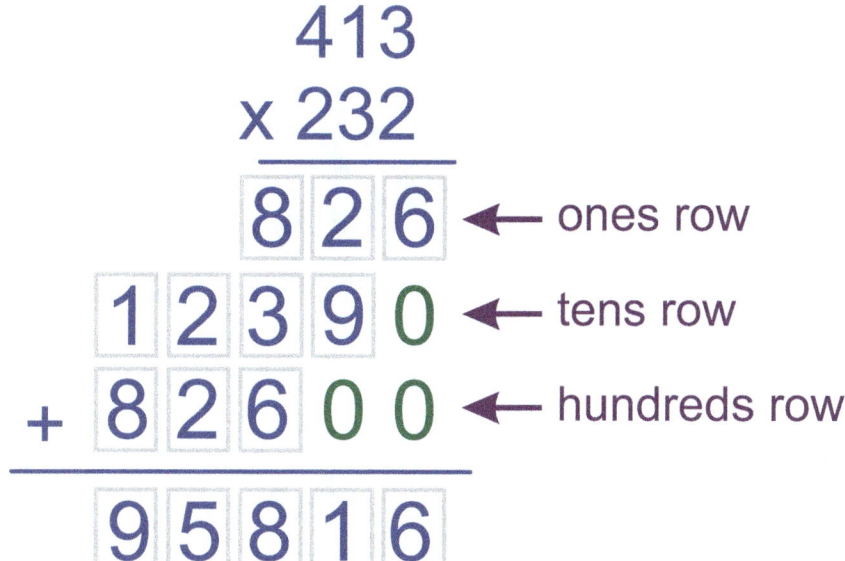

 For every digit of the smaller factor, you should have a row.

Just the Facts! Workbook

Lesson 22: Multiplying Multiple Digits by Three Digit Numbers

Solve the equations below.

(22.1)
```
   123
 x 321
 ─────
   ☐☐☐
  ☐☐☐0
 ☐☐☐00
 ─────
 ☐☐☐☐☐
```

(22.2)
```
   342
 x 283
 ─────
   ☐☐☐
  ☐☐☐0
 ☐☐☐00
 ─────
 ☐☐☐☐☐
```

(22.3)
```
   456
 x 732
 ─────
   ☐☐☐
  ☐☐☐0
 ☐☐☐00
 ─────
 ☐☐☐☐☐
```

(22.4)
```
   698
 x 456
 ─────
   ☐☐☐
  ☐☐☐0
 ☐☐☐00
 ─────
 ☐☐☐☐☐
```

Just the Facts! Workbook

Lesson 22: Multiplying Multiple Digits by Three Digit Numbers

Review: circle the comma name for the indicated comma.

(22.5) 459,000,188 ↑   billion   million   thousand

(22.6) 1,000,126,119 ↑   billion   million   thousand

(22.7) 78,243,992 ↑   billion   million   thousand

(22.8) 1,921 ↑   billion   million   thousand

Circle the number that is the same as the word number.

(22.9) five hundred thirty-three thousand

     (a) 533,000   (b) 533,000,000

(22.10) four hundred twenty-one thousand, six hundred

     (a) 421,000,600   (b) 421,600

(22.11) seventy million, two thousand, fifty

     (a) 70,2000,50   (b) 70,002,050

(22.12) eight hundred fifty-two thousand, thirty-five

     (a) 852,035,000   (b) 852,035

Lesson 22: Multiplying Multiple Digits by Three Digit Numbers

Independent Work - Page 1

Solve the equations below.

1.
   211
   × 102

2.
   342
   × 283

3.
   281
   × 238

4.
   289
   × 121

Just the Facts! Workbook

Lesson 22: Multiplying Multiple Digits by Three Digit Numbers

Independent Work - Page 2

Circle the comma name for the indicated comma.

5. 889,028,000   billion   million   thousand
   ↑ (under the first comma, between 889 and 028)

6. 6,060,000,000   billion   million   thousand
   ↑ (under the first comma, between 6 and 060)

7. 123,600,010   billion   million   thousand
   ↑ (under the first comma, between 123 and 600)

8. 2,312,890   billion   million   thousand
   ↑ (under the first comma, between 2 and 312)

Circle the number that is the same as the word number.

9. twenty-three billion, two hundred thousand, four

   (a) 23,204,000   (b) 23,200,004

10. twenty-three billion, two hundred four thousand

    (a) 23,204,000   (b) 23,200,004

11. one hundred thirteen million, six hundred one

    (a) 100,013,601   (b) 113,000,601

12. thirty-one billion, one million, six hundred thousand, six

    (a) 31,100,600,006   (b) 31,001,600,006

Just the Facts! Workbook

Lesson 23: Greater, Less Than, Equal

To compare numbers we use special symbols.

> The **greater than** symbol is:
>
>
>
> The **less than** symbol is:
>
>

We read these statements left to right, which is why it matters which direction the symbol is facing.

8 > 2, is read as "eight is **greater** than two".

2 < 8, is read as "two is **less than** eight".

To help remember which direction means **greater than**, think of a hungry alligator.

The hungry alligator always prefers a **larger** serving.
The **larger** number always goes where the **big** mouth is.

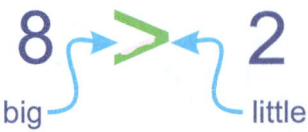

Just the Facts! Workbook

Lesson 23: Greater, Less Than, Equal

Fill in the circles with the symbols: **>** , **<** , or **=** .

(23.1) 7 ◯ 8

(23.2) 56 ◯ 56

(23.3) 100 ◯ 99

(23.4) 12 ◯ 21

(23.5) 512 ◯ 412

(23.6) 421 ◯ 421

If the statement is **incorrect,** cross it out.

(23.7) 88 < 90

(23.8) 76 = 71

(23.9) 542 > 500

(23.10) 33 > 33

(23.11) 611 < 712

(23.12) 1,612 = 1,622

Adding a line under the symbols means that the numbers can *also* be **equal.**

≥ greater or equal

≤ less than or equal

Answer the questions below.

(23.13) Is 44 ≥ 44 ?

yes   no

(23.14) Is 99 ≤ 100 ?

yes   no

Just the Facts! Workbook

Lesson 23: Greater, Less Than, Equal

Review: solve the equations below.

(23.15)

```
    315
  x 452
  ─────
```

(23.16)

```
    681
  x 508
  ─────
```

(23.17)

```
    259
  x 100
  ─────
```

(23.18)

```
   6500
  x 100
  ─────
```

(23.19) (10)(560) = ___ ___ ___ ___

(23.20) 100 * 7020 = ___ ___ ___ ___ ___

Just the Facts! Workbook — 130 —

Lesson 23: Greater, Less Than, Equal

Place value review.

(23.21) Circle the number in the **thousands** spot.

259,321,001

(23.22) Circle the number in the **hundreds** spot.

259,321,001

(23.23) Circle the number in the **tens** spot.

259,321,001

(23.24) Circle the number in the **millions** spot.

259,321,001

(23.25) Circle the number in the **ones** spot.

259,321,001

(23.26) Circle the number in the **ten thousands** spot.

259,321,001

(23.27) Circle the number in the **hundred thousands** spot.

259,321,001

Lesson 23: Greater, Less Than, Equal

Independent Work - Page 1

Use the symbols: **>** , **<** , or **=** , to make the statements true.

1. 80 ◯ 80
2. 32 ◯ 5
3. 245 ◯ 245

4. 54 ◯ 62
5. 145 ◯ 452
6. 890 ◯ 800

Are the statements below true?

7. 54 ⩾ 44
yes   no

8. 25 ⩽ 25
yes   no

9. 61 ⩽ 31
yes   no

---

10. Jill has <u>four hundred twenty-one dollars</u> saved up. Her friend, Emma, has <u>four hundred twenty dollars</u> saved up.
Is the following statement **true** or **false**?

Does Jill has more money than Emma? _____

Write the equation that proves your answer.

[   ] ◯ [   ]

Just the Facts! Workbook   - 132 -

Lesson 23: Greater, Less Than, Equal

Independent Work - Page 2

Solve the equations below.

11.
```
   23
x  10
```

12.
```
  235
x 100
```

13.
```
 1000
x  30
```

14.
```
   44
x  14
```

15.
```
   68
x  23
```

16.
```
   6 2 3
x  1 0 6
```

17.
```
   3 4 0
x  1 2 4
```

- 133 -

Lesson 24: Rounding Numbers

Sometimes you don't need an exact number, and it is helpful to round a large number to the nearest ten, hundred, thousand, etc.

How to Round a Number

Step 1. Underline the number in the spot to round to.

Example: Round 532 and 536 to the nearest 10.

First, underline the **tens** number.

5<u>3</u>2   5<u>3</u>6

— This is the tens spot. —

Step 2. If the number to the **right** of the underlined number is greater or equal to 5, add one to the underlined number. Otherwise, do not change it.

5<u>3</u>2 → 5<u>3</u>2   5<u>3</u>6 → 5<u>4</u>6

2 is NOT ≥ 5, leave the 3

6 ≥ 5, add 1 to the 3 to get 5<u>4</u>6

Step 3. Change all numbers, to the **right** of the underlined number to **zeros.**

5<u>3</u>2 → 5<u>3</u>0   5<u>4</u>6 → 5<u>4</u>0

<u>532</u> rounded to the nearest 10 is <u>530</u>.

<u>536</u> rounded to the nearest 10 is <u>540</u>.

Just the Facts! Workbook

Lesson 24: Rounding Numbers

The numbers below are rounded to the nearest 10.

	step 1	step 2	step 3
43	4<u>3</u>	3 is NOT ≥ 5, keep the 4	4<u>0</u>
49	4<u>9</u>	9 is ≥ 5, add 1 to the 4	5<u>0</u>
45	4<u>5</u>	5 is ≥ 5, add 1 to the 4	5<u>0</u>

The numbers below are rounded to the nearest 100.

	step 1	step 2	step 3
263	2<u>6</u>3	6 is ≥ 5, add 1 to the 2	3<u>00</u>
249	2<u>4</u>9	4 is NOT ≥ 5, keep the 2	2<u>00</u>
250	2<u>5</u>0	5 is ≥ 5, add 1 to the 2	3<u>00</u>

(24.1) Round 282 to the nearest ten. _____

(24.2) Round 120 to the nearest hundred. _____

(24.3) Round 47 to the nearest ten. _____

(24.4) Round 653 to the nearest hundred. _____

Lesson 24: Rounding Numbers

Adding a Digit When Rounding

Sometimes you have to add another digit to the number when rounding. For example, 970 rounded to the nearest 100 would be 1000. Making the number four digits instead of three.

Round 98 to the nearest ten.

Step 1: under line the number in the 10 spot.   9̲8

Step 2: 8 ≥ 5 so add one to the 9 to make it 10.   1̲0̲8

Step 3: make numbers to the right of underlined number 0.   1̲0̲0

9̲8̲ rounded to the nearest ten is 1̲0̲0̲.

(24.5) Round 98 to the nearest ten. _____

(24.6) Round 960 to the nearest hundred. _____

(24.7) Round 6,700 to the nearest thousand. _____

(24.8) Round 22,950 to the nearest ten thousand. _____

Find the products and then use the symbols.
>, <, or =   to make the statements below true.

(24.9)   7(6) ◯ 42

(24.10)   7•7 ◯ 50

(24.11)   9 * 9 ◯ 100

(24.12)   8 x 8 ◯ 35

Just the Facts! Workbook

Lesson 24: Rounding Numbers

Independent Work - Page 1

Round the numbers below to the nearest **ten**.

1. 67 ____
2. 85 ____
3. 92 ____
4. 238 ____
5. 822 ____
6. 655 ____
7. 94 ____
8. 96 ____
9. 95 ____

Round the numbers below to the nearest **hundred**.

10. 670 ____
11. 859 ____
12. 811 ____
13. 920 ____
14. 981 ____
15. 950 ____

Round the numbers below to the nearest **thousand**.

16. 3,299 ____
17. 43,612 ____
18. 9,813 ____
19. 8,500 ____

Lesson 24: Rounding Numbers

Independent Work - Page 2

Find the products and then use the symbols.
$>$, $<$, or $=$ to make the statements below true.

20. 7(8) ◯ 100

21. 9•9 ◯ 80

22. 5 * 6 ◯ 20

23. 2 x 9 ◯ 15

Solve the equations below.

24.
```
  52
x 38
```

25.
```
  81
x 76
```

---

26. Tom has <u>ten times</u> more pennies in a jar than Kim has. Kim has <u>six-two thousand, one hundred two</u> pennies. How many pennies does Tom have? Write the answer using numbers.

Just the Facts! Workbook

Lesson 25: Multiplying Multiples of Ten

When your multiplying two numbers and one or both is a multiple of ten, you can make the problem easier to solve. You may even be able to solve it in your head, without having to write it out.

$$\begin{array}{r} 300 \\ \times\ 4 \\ \hline \end{array}$$

3̶0̶0̶ → take away the zeros so 300 becomes 3 → 3
× 4   × 4

3   add the zeros that
× 4   you took away   →   1200
―――
 12

solve the
new problem

---

300   take the 3   3   add the
× 40   zeros away   →   × 4   → 3 zeros   →   12,000
                          ―――
                           12

1200   take the 4   12   add the
× 200 zeros away   →   × 2   → 4 zeros   →   240,000
                          ―――
                           24

Lesson 25: Multiplying Multiples of Ten

(25.1)

400
x   8

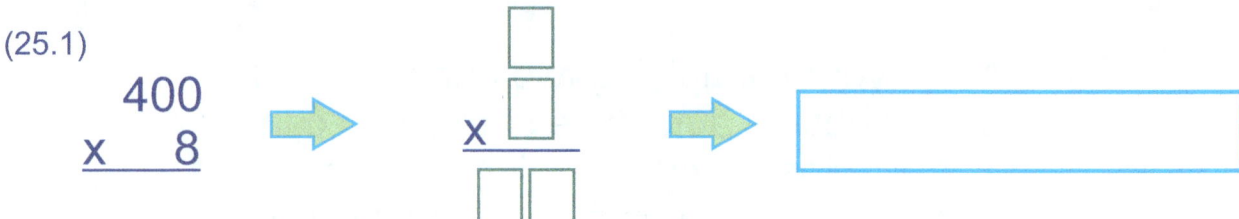

(25.2)

50
x 50

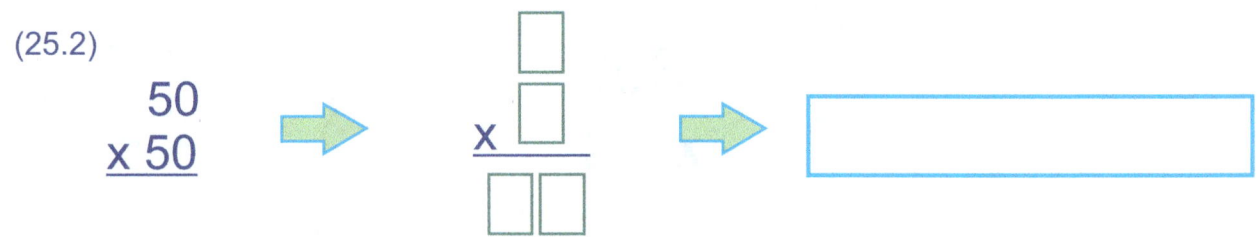

(25.3)

3000
x 250

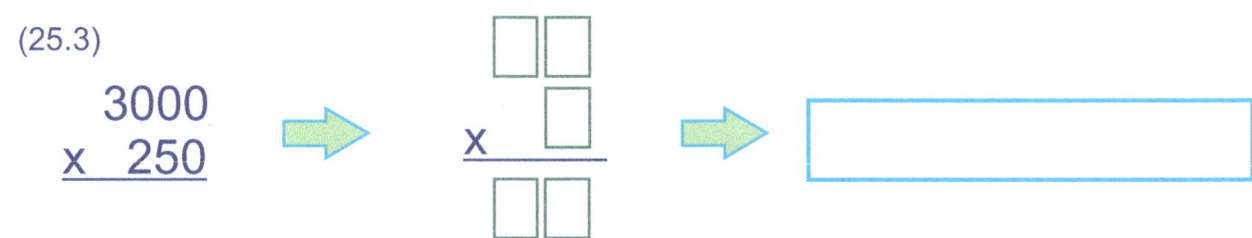

(25.4)

12000
x   300

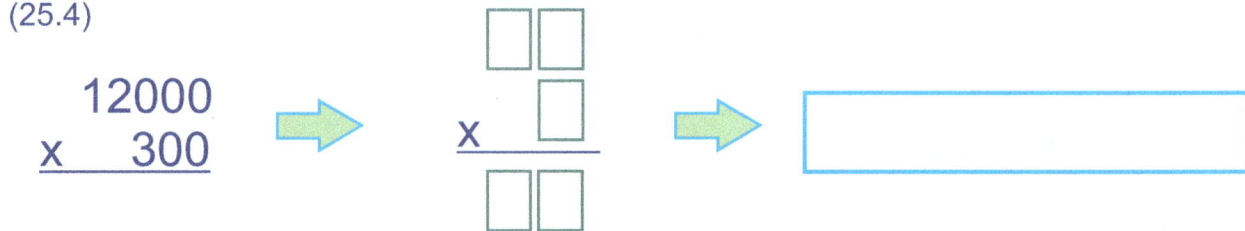

(25.5)

9210
x 200

Lesson 25: Multiplying Multiples of Ten

Rounding Numbers Revisited

When rounding a number like 6,983 to the nearest 100, the nine gets a one added to it, so it's now a 10. Now you must carry the one to the number to the left of the underlined number.

Round 6,983 to the nearest hundred.

6,9̲83   8 ≥ 5, so add 1 to the 9 = 1̲0, carry the 1 → 6,0̲83

    1
6,0̲83 → 70̲83 → 70̲00

Round the numbers below to the nearest **hundred.**

(25.6) 992 [ ]

(25.7) 1923 [ ]

(25.8) 25,971 [ ]

(25.9) 87,860 [ ]

(25.10) 80,928 [ ]

(25.11) 60,950 [ ]

(25.12)

Erin walks <u>eight thousand</u> steps every day. If the month has <u>thirty</u> days, how many steps does Erin walk in the month? Write the answer using numbers.

Lesson 25: Multiplying Multiples of Ten

Independent Work - Page 1

Find the products and then use the symbols
>, <, or = to make the statements below true.

1. 4 x 8 ◯ 3 x 3

2. 3•6 ◯ 2•9

3. 3(5) ◯ 8(2)

4. 12 x 2 ◯ 8 x 3

Solve the equations below.

5. 900
   x 8
   ⇒  ⇒

6. 90
   x 50
   ⇒  ⇒

7. 2500
   x   30
   ⇒ 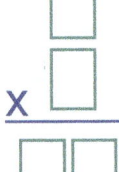 ⇒

Just the Facts! Workbook

Lesson 25: Multiplying Multiples of Ten

Independent Work - Page 2

Round the numbers below to the nearest **hundred**.

8. 965 [ ]

9. 2984 [ ]

10. 3919 [ ]

11. 23,981 [ ]

12. 855 [ ]

13. 950 [ ]

14. Round the number to the nearest **ten thousand**.

   78,135 [ ]

15. Round the number to the nearest **hundred thousand**.

   136,457 [ ]

Solve the equations.

16.
$$\begin{array}{r} 37 \\ \times 52 \\ \hline \end{array}$$
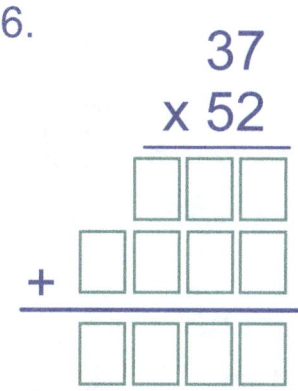

17.
$$\begin{array}{r} 91 \\ \times 62 \\ \hline \end{array}$$
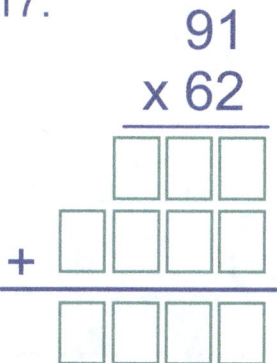

Just the Facts! Workbook

Lesson 26: Estimating

Sometimes you don't need to find an exact answer and can estimate. Estimating is helpful, for example, if you're using a calculator. If you enter in a wrong number, but have an idea of what the answer should be, then you have a chance at catching the error. Estimating is also useful when the exact answer is not needed and you only need an idea.

> **How to Estimate Multiplication Problems**
>
> Change the numbers so that they are multiples of 10, and then multiply.

Lauren has <u>53</u> buckets of apples. Each bucket has <u>68</u> apples. **About** how many apples does Lauren have?

Step 1: Round 53 to 50
Step 2: Round 68 to 70
Step 3: Multiply 50 x 70 = <u>3,500 apples</u>

Last year John biked 18 miles 216 times. **About** how many miles did John bike for the year?

Step 1: Round 18 to 20
Step 2: Round 216 to 220
Step 3: Multiply 20 x 220 = <u>4,400 miles</u>

> **Words That Tell You to Estimate**
>
> "about how many", "approximately", or "approximate"

Just the Facts! Workbook

Lesson 26: Estimating

Solve the problems by estimating.

(26.1) 41 x 39 = ➡ ____ x ____ = [   ]

(26.2) 98 x 24 = ➡ ____ x ____ = [   ]

(26.3) 55 x 63 = ➡ ____ x ____ = [   ]

(26.4) Jack has 23 bags of beans. Each bag has 38 beans. **Approximately**, how many beans does Jack have?

____ x ____ = ➡ ____ x ____ = [   ]

(26.5) The baseball league had 22 teams. There are 18 players on each team. Every player needs to have a shirt. **About** how many shirts will the league need to order?

____ x ____ = ➡ ____ x ____ = [   ]

Lesson 26: Estimating

Find the products and then use the symbols
>, < , or = to make the statements below true.

(26.6) 7 * 8 ◯ 11*5

(26.7) 5 x 8 ◯ 10(4)

(26.8) (9)(7) ◯ 9 * 7

(26.9) 8(3) ◯ 12(2)

Solve the equations below.

(26.10) 600 x 6 ➡ ☐ x ☐ / ☐☐ ➡ ▭

(26.11) 700 x 50 ➡ ☐ x ☐ / ☐☐ ➡ ▭

(26.12) 3100 x 70 ➡ ☐☐ x ☐ / ☐☐☐ ➡ ▭

Just the Facts! Workbook

Lesson 26: Estimating

Independent Work - Page 1

Solve the problems by estimating.

1. 32 x 32 = ➡ ____ x ____ = [_____]

2. 68 x 31 = ➡ ____ x ____ = [_____]

3. 43 x 37 = ➡ ____ x ____ = [_____]

---

4. There are 88 people invited to the party. Each person needs at least two napkins. **About** how many napkins should be bought?

____ x ____ = ➡ ____ x ____ = [_____]

5. The marathon has exactly 478 people attending. Each person needs to fill out 9 forms. **About** how many forms need to be provided?

____ x ____ = ➡ ____ x ____ = [_____]

Just the Facts! Workbook

Lesson 26: Estimating

Independent Work - Page 2

Circle the value of the underlined digit.

6. <u>3</u>30,001   hundred thousands   ten millions

7. 5<u>6</u>8,012   ten thousands   thousands

8. <u>8</u>,012,899   billions   millions

9. <u>9</u>9,778   ten millions   ten thousands

10. 1<u>2</u>,189   ten thousands   thousands

11. Solve the problem and answer the question.

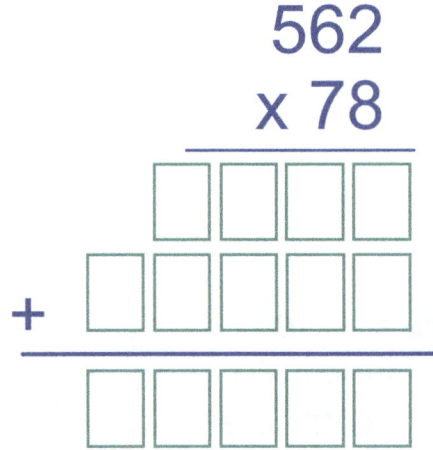

$$562 \times 78$$

Which number, in the answer above, is in the <u>ten thousands</u> spot?

Just the Facts! Workbook

Lesson 27: Expanding Numbers

A large, **standard number** can be **expanded** to an addition problem created out of numbers that are multiples of their place values. This is helpful for solving problems, and will eventually turn into scientific notation for later math and science studies.

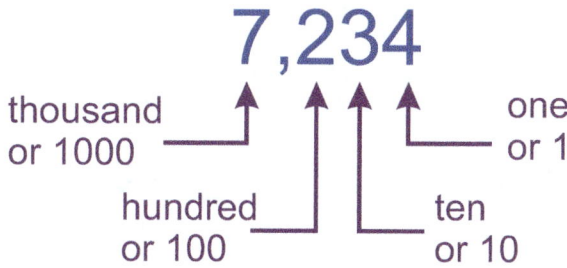

$$7,234 = 7,000 + 200 + 30 + 4$$

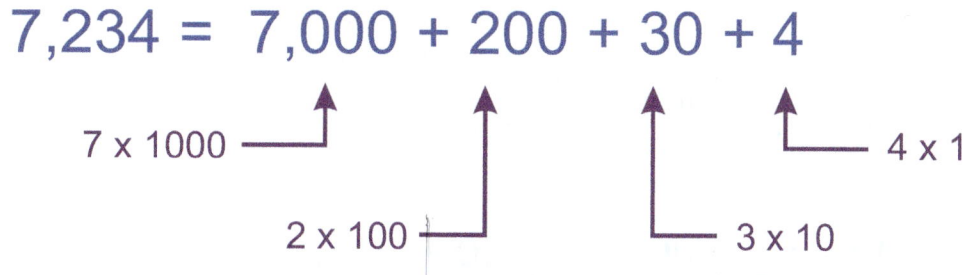

Notice how each digit in the number is in the equation as a multiple of its place value, and when you add them all up (see below) you get the original un-expanded number.

```
  7000
   200
    30
+    4
------
  7234
```

(27.1) Fill in the blanks to expand the number.

25,896 = ___0,000 + ___,000 + ___00 + ___0 + ___

Just the Facts! Workbook

Lesson 27: Expanding Numbers

If a digit in the large number is a zero, you can skip that term in the expanded number.

20,856 = __2__0,000 + __8__00 + __5__0 + __6__

↑ Nothing for the thousands (1000) place.

Fill in the blanks to expand the numbers below.

(27.2) 92,341 = ___0,000 + ___,000 + ___00 + ___0 + ___

(27.3) 168,520 = ___00,000 + ___0,000 + ___,000 + ___00 + ___0

(27.4) 6030 = ___000 + ___0

(27.5) 5,432 = ___000 + ___00 + ___0 + ___

Standard numbers are the opposite of expanded numbers.

Un-expand the numbers below, so that they are in standard form.

(27.6) 40,000 + 1,000 + 800 + 50 + 2 = _____

(27.7) 60,000 + 8,000 + 900 = _____

(27.8) 800,000 + 60,000 + 4,000 + 800 = _____

Lesson 27: Expanding Numbers

Independent Work - Page 1

Fill in the blanks to expand the numbers below.

1. 13,679 = ___0,000 + ___,000 + ___00 + ___0 + ___

2. 4,310 = ___000 + ___00 + ___0

3. 231,520 = ___00,000 + ___0,000 + ___,000 + ___00 + ___0

4. 25,386 = ___0,000 + ___000 + ___00 + ___0 + ___

Expand the numbers below.

5. 684 = _____

6. 1459 = _____

7. 925 = _____

Un-expand the numbers below, so that they are in **standard** form.

8. 60,000 + 2,000 + 700 + 30 + 9 = _____

9. 50,000 + 8,000 + 400 = _____

Just the Facts! Workbook

Lesson 27: Expanding Numbers

Independent Work - Page 2

Solve the equations below.

10. 400
    x 30 ➡ ☐
          x ☐   ➡ ☐
          ——
          ☐☐

11. 80
    x 70   ➡ ☐
          x ☐   ➡ ☐
          ——
          ☐☐

12. 4000
    x   80 ➡ ☐
          x ☐   ➡ ☐
          ——
          ☐☐

Solve the problems by estimating.

13. 105 x 23 = ➡ ____ x ____ = ☐

14. 55 x 78 = ➡ ____ x ____ = ☐

15. 18 x 11 = ➡ ____ x ____ = ☐

Lesson 28 — Multiplying with the Box Method

Another way to multiply two numbers, is to use the **box method**, which is also known as the **partial products method**.

Step 1 - Create a table where the numbers of rows is equal to the number of digits in one factor, and the number of columns are equal to the number of digits in the other factor.

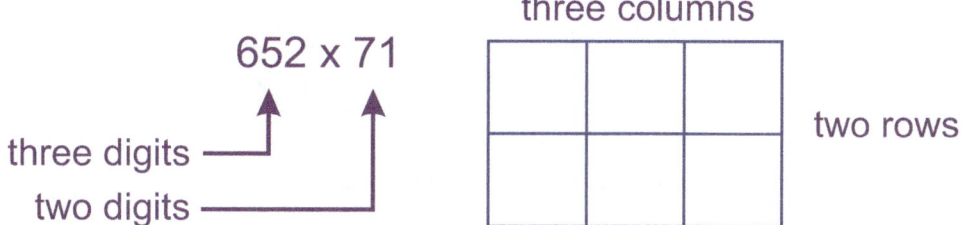

Step 2 - Expand the factors, putting those terms outside the box as done below.

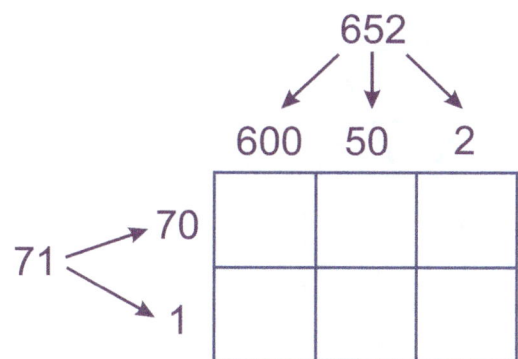

Step 4 - Fill in the table by **multiplying** the rows and columns.

	600	50	2
70	42,000	3,500	140
1	600	50	2

Just the Facts! Workbook

Lesson 28: Multiplying with the Box Method

Step 5 - Add **all** of the products in the table.

	600	50	2
70	42,000	3,500	140
1	600	50	2

```
  42000
   3500
    140
    600
     50
+     2
  46292
```

652 x 71 = 46,292

Let's check our answers using the standard way of multiplying.

(28.1)

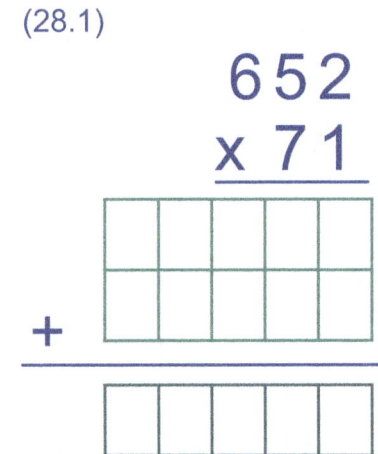

The box method helps us see how the numbers are related. Also, for smaller factors, we could more easily do this in our heads.

25 x 8

	20	5
8	160	40

160 + 40 = **200**

Just the Facts! Workbook

Lesson 28: Multiplying with the Box Method

Use the box method to solve the equations.

(28.2)

42 x 86

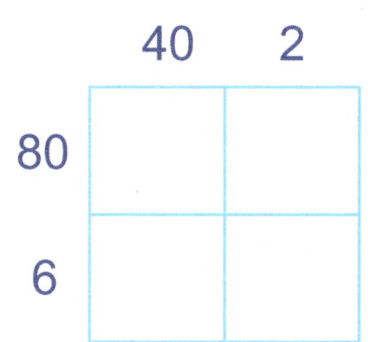

42 x 86 = _____

(28.3)

524 x 38

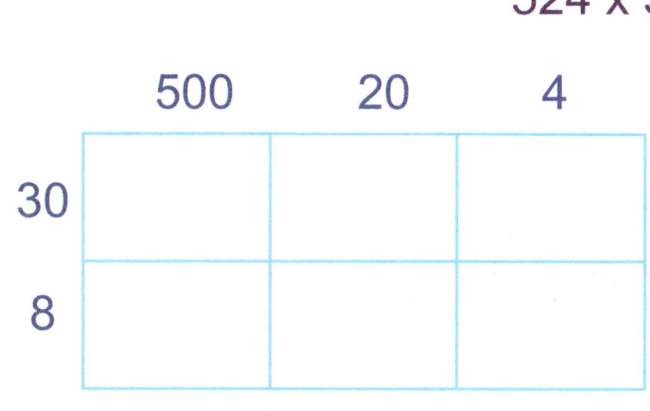

524 x 38 = _____

Lesson 28: Multiplying with the Box Method

Fill in the blanks for the equations below.

(28.4) 6 x _____ = 36

(28.7) 7 x _____ = 49

(28.5) 7 x _____ = 21

(28.8) 6 x _____ = 42

(28.6) 8 x _____ = 72

(28.9) 8 x _____ = 64

Expand the numbers below.

(28.10)

8,752,005 = _____

_____

(28.11)

95,100,780 = _____

_____

(28.12) Circle the **factors** and underline the **products**,

62 x 3 = 186

11 x 40 = 440

7 x 20 = 140

160 x 20 = 3,200

Just the Facts! Workbook

Lesson 28: Multiplying with the Box Method

Independent Work - Page 1

Use the box method to solve the equations.

1.

56 x 39

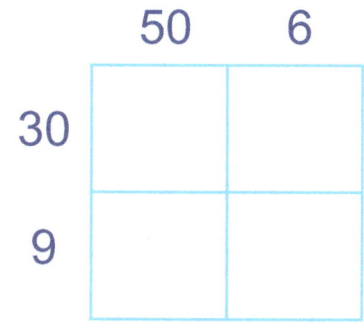 

56 x 39 = _____

2.

15 x 74

15 x 74 = _____

Lesson 28: Multiplying with the Box Method

Independent Work - Page 2

Fill in the blanks for the equations below.

3. 9 x ____ = 81   8. 3 x ____ = 27

4. 3 x ____ = 24   9. 5 x ____ = 15

5. 4 x ____ = 20   10. 2 x ____ = 12

6. 6 x ____ = 30   11. 4 x ____ = 12

7. 7 x ____ = 42   12. 3 x ____ = 15

Circle the value of the underlined digit.

13. 3<u>5</u>4,787   thousands   ten thousands

14. <u>5</u>76,274   thousands   hundred thousands

Answer the questions below.

15. 5,412   What number is in the <u>tens</u> spot? _____

16. 63,194   What number is in the <u>ten thousands</u> spot? _____

Just the Facts! Workbook

Lesson 29: The Distributive Property

The **distributive property** is very important. It shows how an equation can be changed to make it easier to solve.

You can use this property if you have a sum (or difference) of two or more numbers that is multiplied by another number.

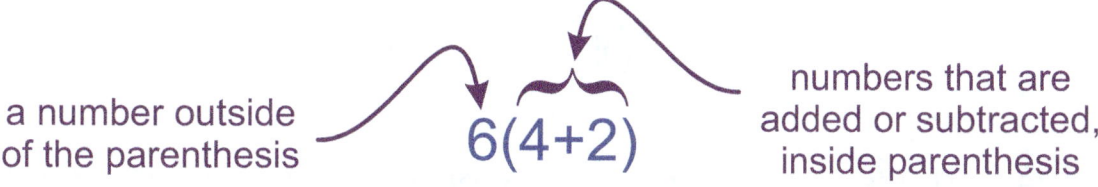

a number outside of the parenthesis — 6(4+2) — numbers that are added or subtracted, inside parenthesis

We "distribute" the 6, by multiplying it by the numbers inside the parenthesis as done below.

$$6(4+2) = 6(4) + 6(2)$$
$$= 24 + 12$$
$$= \underline{36}$$

We can check this by adding the numbers in the parentheses, and then multiplying. You will see that the products are equal.

$$6(4+2) = 6(6) = \underline{36}$$

* This property will be very important when you learn algebra, when numbers are substituted for letters (variables).

Just the Facts! Workbook

Lesson 29: The Distributive Property

Why is the distributive property useful?
Sometimes it's easier to do an equation a different way than it's written. For example, 7 x 13 (see below) is easier to do in your head if you approach it a different way.

$$7 \times 13$$

Expand the 13
13 = 10 + 3

$$7(10 + 3)$$

Now we can use the **distributive property** to solve the equation.

$$7(10 + 3)$$

$$7(10) + 7(3)$$

$$70 + 21$$

$$= 91$$

Checking the equation using the carry method, we can see that the products are equal.

```
  2
 13
x 7
 ---
 91
```

Just the Facts! Workbook

Lesson 29: The Distributive Property

Solve the following two ways.
1) Solve using the distributive property.
2) Solve by first doing the addition inside the parenthesis.

(28.1)
5(4 + 3)

5(4 + 3)

____ + ____

= ____

5(4 + 3)

4 + 3 = ____

5 x ____ = ____

(28.2)
2(3 + 5)

2(3 + 5)

____ + ____

= ____

2(3 + 5)

3 + 5 = ____

2 x ____ = ____

(28.3)
6(7 + 2)

6(7 + 2)

____ + ____

= ____

6(7 + 2)

7 + 2 = ____

6 x ____ = ____

Lesson 29: The Distributive Property

Independent Work - Page 1

Solve the following two ways.
1) Solve using the distributive property.
2) Solve by first doing the addition inside the parenthesis.

1.
$$4(3 + 2)$$

$4(3 + 2)$ $4(3 + 2)$

___ + ___ $3 + 2 =$ ___

= ___ $4 \times$ ___ = ___

2.
$$2(2 + 8)$$

$2(2 + 8)$ $2(2 + 8)$

___ + ___ $2 + 8 =$ ___

= ___ $4 \times$ ___ = ___

3.
$$5(5 + 6)$$

$5(5 + 6)$ $5(5 + 6)$

___ + ___ $5 + 6 =$ ___

= ___ $5 \times$ ___ = ___

Just the Facts! Workbook

Lesson 29: The Distributive Property

Independent Work - Page 2

4. Use the box method to solve: 15 x 27

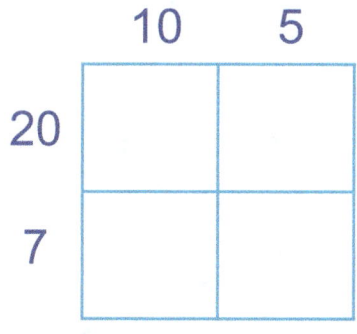
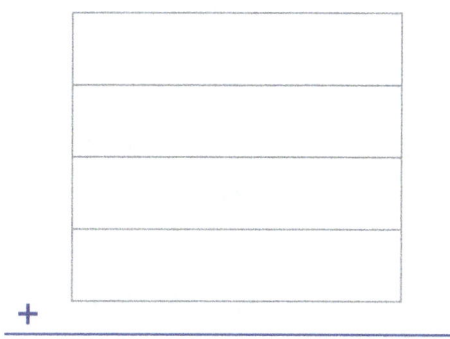

15 x 27 = _____

5. Solve the equations below.

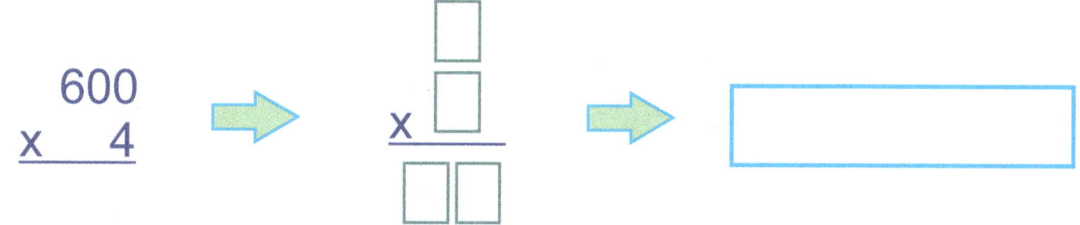

6. Round 66 to the nearest 10.   _____

7. Round 782 to the nearest 100.   _____

8. Round 55 to the nearest 10.   _____

9.   4(10) = _____   12. 8 x _____ = 64

10. 3 x 10 = _____   13. 3 x _____ = 9

11. 35 * 100 = _____   14. 9 x _____ = 54

Lesson 30: Introduction to Exponents

Equations where numbers are multiplied over and over can be written in **exponent** form.

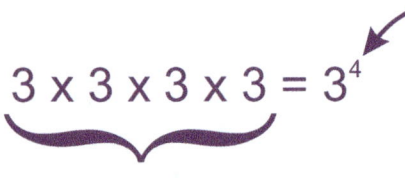

$3 \times 3 \times 3 \times 3 = 3^4$ — this small number is an **exponent**. It tells us how many times the other number, the *base*, is multiplied by.

Four threes are multiplied.

Write the equations in **exponent** form.

(30.1) $5 \times 5 \times 5 =$ _____

(30.2) $6 \times 6 =$ _____

(30.3) $7 \times 7 \times 7 \times 7 =$ _____

(30.4) $6 \times 6 \times 6 \times 6 \times 6 =$ _____

Change the equations from **exponent** form to standard form.

(30.5) $2^4 =$ _____

(30.6) $6^3 =$ _____

(30.7) $7^5 =$ _____

(30.8) $9^2 =$ _____

Just the Facts! Workbook

Lesson 30: Introduction to Exponents

If the exponent is one, the base number is multiplied one time.

$$3^1 = 3$$

3, one time, is three.

> When the **exponent is zero**, the answer is ALWAYS 1.
>
> $$3^0 = 1$$

To solve an equation with an exponent,
multiply two numbers at a time, as done below.

$$2^4 = 2 \times 2 \times 2 \times 2$$
$$4$$
$$4 \times 2 \times 2$$
$$8$$
$$8 \times 2$$
$$= 16$$

$$6^3 = 6 \times 6 \times 6$$
$$36$$
$$36 \times 6$$
$$= 216$$

$$\begin{array}{r} \overset{3}{3}6 \\ \times\ 6 \\ \hline 216 \end{array}$$

Just the Facts! Workbook

Lesson 30: Introduction to Exponents

Solve the following equations.

(30.9) $2^0 = $ _____

(30.10) $6^1 = $ _____

(30.11) $7^2 = $ _____

(30.12) $9^2 = $ _____

(30.13) $2^3 = $ ___ x ___ x ___

___ x ___

= _____

Use the symbols: **>** , **<** , or **=** to make the statements below true.

(30.14) 2^2 ◯ 12

(30.15) 6^2 ◯ 37

(30.16) 8^0 ◯ 0

(30.17) 7^2 ◯ 49

(30.18) 9^1 ◯ 9

(30.19) 10^2 ◯ 100

Lesson 30: Introduction to Exponents

Independent Work - Page 1

Circle the equation that is equal to the one on the left.

1. $8^3 =$ 8 x 8 x 8   8 + 8 + 8   8 x 8

2. $4^2 =$ 4 x 4 x 4   4 x 4   4 + 4

3. $6^3 =$ (6)(6)(6)   6 x 6 x 6 x 6   6+6+6

4. $7^4 =$ 7 x 7   7 x 7 x 7   7 x 7 x 7 x 7

5. $9^2 =$ 9 * 9   9 * 9 * 9   9 + 9

Solve the equations below.

6. $3^1 =$ _____   8. $4^1 =$ _____   10. $28^1 =$ _____

7. $2^0 =$ _____   9. $9^0 =$ _____   11. $36^0 =$ _____

12. Henry went to the store and bought 5 packs of pencils; each pack has 5 pencils. Which equation below will equal the total number of pencils? Hint: there are ____ groups of ____ pencils.

(a) 5 x 5 x 5   (b) 5^5 (c) 5^2

Just the Facts! Workbook

Lesson 30: Introduction to Exponents

Independent Work - Page 2

Fill in the blanks for the equations below.

13. 2 x _____ = 12

14. 3 x _____ = 15

15. 5 x _____ = 30

16. 6 x _____ = 36

17. 8 x _____ = 16

18. 3 x _____ = 12

19. 5 x _____ = 25

20. 6 x _____ = 42

21. 7 x _____ = 56

22. 8 x _____ = 24

---

23. There are 365 days in a year. How many days are in 10 years (not including leap years).

---

24. Circle the equation that has the same answer (product) as the equation below.

$$4 \times 59$$

(a) 4(50 x 9)   (b) 4(50 + 9)   (c) 4(50 - 9)

Lesson 31: Division

When you divide something, you're breaking it up into equal smaller pieces.

In the fish tank above, we have 12 fish.
If you divide the fish into 4 **equal** groups, you have:

$$12 \div 4 = 3$$

This is read as, "Twelve divided by four equals three".

12 fish, divided into 4 equal groups, equals 3 fish for each group.

The number being divided is called the dividend.

The answer is called the quotient.

$$12 \div 4 = 3$$

The number of groups being divided into is called the divisor.

Lesson 31: Division

> There are **three** ways to write a division problem.
>
> ① $12 \div 4 = 3$ ② $\dfrac{12}{4} = 3$ ③ $4\overline{)12}^3$

① The **first** number (the dividend) is the one that gets chopped up.

 $15 \div 5 = 3$

② Here, you can think that the number, 15 (the dividend), is *in*side the **chopping house**.

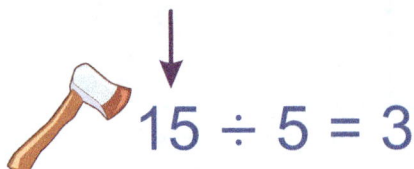

③ Here, the 15 (the dividend), *is on top* of the **chopping block**.

Reminder: the answers for addition, subtraction, multiplication and division, all have their own names.

$5 + 3 = 8$ — sum $5 - 3 = 2$ — difference $5 \times 3 = 15$ — product $15 \div 3 = 5$ — quotient

Just the Facts! Workbook

Lesson 31: Division

(31.1) We have ten apples, and want to divide them by two. Circle the apples above to make two equal groups.

Ten divided by two is equal to _____.

10 ÷ 2 = ____ $\frac{10}{2}$ = ____   2)⎺1⎺0⎺

(31.2) We have sixteen cupcakes, and want to divide them by four. Circle the cupcakes above to make four equal groups.

Sixteen divided by four is equal to _____.

16 ÷ 4 = ____ $\frac{16}{4}$ = ____   4)⎺1⎺6⎺

Just the Facts! Workbook

Lesson 31: Division

Did you notice that division is the opposite of multiplication? In fact, you can check your answers by multiplying.

①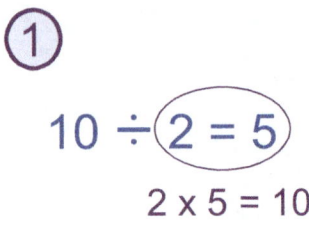

$10 \div 2 = 5$

$2 \times 5 = 10$

②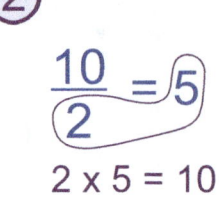

$\frac{10}{2} = 5$

$2 \times 5 = 10$

③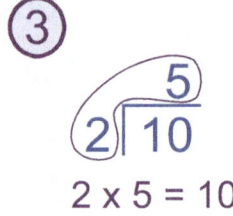

$2\overline{)10}$ with quotient 5

$2 \times 5 = 10$

(31.3) Circle the two numbers to multiply to check the answer.

$16 \div 2 = 8$ $\frac{16}{2} = 8$ $2\overline{)16}$ with quotient 8

Knowing that we can use multiplication to check our answers, we know that we can use multiplication to **find** the answer.

(31.4) $20 \div 4 = $ _____   ⇨ $4 \times ? = 20$

(31.5) $80 \div 8 = $ _____   ⇨ $8 \times ? = 80$

(31.6) $45 \div 5 = $ _____   ⇨ $5 \times ? = 45$

(31.7) $33 \div 3 = $ _____   ⇨ $3 \times ? = 33$

Just the Facts! Workbook

Lesson 31: Division

Solve the division problems below.

(31.8) $99 \div 11 = ?$ ⇒ ____ x ____ = 99   ☐

(31.9) $54 \div 9 = ?$ ⇒ ____ x ____ = 54   ☐

(31.10) $30 \div 5 = ?$ ⇒ ____ x ____ = 30   ☐

(31.11) $6\overline{)60}$ ⇒ ____ x ____ = 60   ☐

(31.12) $9\overline{)27}$ ⇒ ____ x ____ = 32   ☐

(31.13) $\dfrac{42}{6} = ?$ ⇒ ____ x ____ = 42   ☐

(31.14) $\dfrac{49}{7} = ?$ ⇒ ____ x ____ = 49   ☐

Lesson 31: Division

Independent Work - Page 1

1. We have fifteen apples, and want to divide them by three. Circle the apples above to make three equal groups.

Fifteen divided by three is _____.

2. Fill in the blanks to show the equations for the statement from #1 above: Fifteen divided by three is five.

$$\frac{\Box}{\Box} = \Box$$

$$\Box \overline{)\Box}$$

3. What is the **Quotient** in the above equations? ☐

Just the Facts! Workbook

Lesson 31: Division

Independent Work - Page 2

Write the division problem **three** ways.

4. Thirty-six divided by two

5. Twenty-one divided by seven

6. Fifty-five divided by eleven

Solve the division problems below.

7. 9 ÷ 3 = ? ➡ 3 x ? = 9

8. 24 ÷ 8 = ? ➡ 8 x ? = 24

9. 18 ÷ 3 = ? ➡ 3 x ? = 18

Lesson 32

Word Problems

Word problems can be scary. What you need to do is break them down into an equation that you can solve. Try to get rid of all the extra words that you don't need, get the information you do need. Also, it may help to draw out what the problem is asking so you have a better way to understand it.

Step 1: Underline the numbers in the word problem.

Step 2: Underline key words that tell you what to do. These words can be straight forward like "divide", or they can be subtle, like "separate equally". Other words to look for: "in all", "more than", "less than", "times", etc.

Step 3: Determine what kind of problem this is; do you need to add, subtract, multiply or divide? Then write out the equation.

Step 4: Solve the equation.

(32.1) Jim has <u>fifteen dollars</u>, and he wants to pay his <u>three children</u> for raking the leaves. Since each child does the same amount of work, Jim wants to <u>divide</u> the money equally. How much does Jim pay each child?

Lesson 32: Word Problems

Circle the equation that goes with the word problem.

(32.2)

Elmsville Day Camp has three hundred campers. If they have ten rooms, and want to separated the campers in the rooms (equally), how many campers will be assigned to each room?

(a) 300 ÷ 10 =   (b) 300 x 10 =   (c) 300 - 10 =

(32.3)

If you baked twenty-one cookies, and wanted to pass them out (equally) to yourself *and* six other friends, how many cookies does each friend get?

(a) 21 ÷ 6 =   (b) 21 ÷ 7 =   (c) 21 - 6 =

(32.4)

If you baked twenty-one cookies, but you need fifty, how many more cookies do you need to bake?

(a) 21 ÷ 50 =   (b) 50 ÷ 21 =   (c) 50 - 21 =

Solve equations below.

(32.5)

7)49

7 x ? = 49

(32.6)

$\frac{20}{4} = $ ____

4 x ? = 20

(32.7)

50 ÷ 5 = ____

5 x ? = 50

Just the Facts! Workbook

Lesson 32: Word Problems

Circle the equations that the equal the equation on the left.

(32.8) 6^2	(a) 6 x 2	(b) 6 * 6	(c) 6 x 6
(32.9) 7^2	(a) 7 * 7	(b) (7)(7)	(c) 2(7)
(32.10) 3^3	(a) 3 + 3 + 3	(b) 3(3)(3)	(c) 3 x 3
(32.11) 2^3	(a) 2 * 3	(b) (2)(2)(2)	(c) 2 * 2 * 2

(32.12)

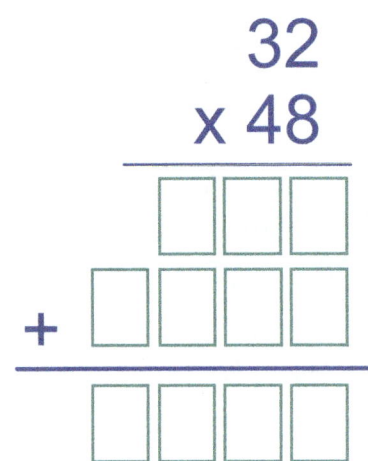

(32.13)

$$\begin{array}{r} 50 \\ \times\ 6 \\ \hline \end{array}$$

(32.14)

$$\begin{array}{r} 900 \\ \times\ \ 8 \\ \hline \end{array}$$

(32.15)

$$\begin{array}{r} 40 \\ \times\ 60 \\ \hline \end{array}$$

(32.16) What is the answer to a division problem called?

(a) product   (b) sum   (c) quotient

Lesson 32: Word Problems

Independent Work - Page 1

Circle the equation that goes with the word problem.

1. Jim rides his bike every day for two hours. How many hours does Jim ride his bike in the month of July. Note that July has 31 days.

   (a) 31 ÷ 2 =   (b) 2 x 31 =   (c) 31 - 2 =

2. There are twelve eggs in a dozen. Jack bought 3 dozen cartons of eggs. How many eggs does Jack have?

   (a) 12 ÷ 3 =   (b) 12 x 3 =   (c) 12 + 3 =

3. You invited three friends to go to the fair with you. You plan on giving each friend ten tickets. How many tickets should you purchase for you and your friends?

   (a) 10 x 3 =   (b) 10 x (3+1) =   (c) 10 + (3+1) =

4.
```
   60
 x  6
```

5.
```
  700
 x  7
```

6.
```
   90
 x 90
```

Just the Facts! Workbook

Lesson 32: Word Problems

Independent Work - Page 2

Circle the names for the underlined numbers.

		sum	product	quotient	factor
7.	5 x 6 = <u>30</u>	sum	product	quotient	factor
8.	30 ÷ 6 = <u>5</u>	sum	product	quotient	factor
9.	6 + 5 = <u>11</u>	sum	product	quotient	factor
10.	<u>6</u>(5) = 30	sum	product	quotient	factor

11.

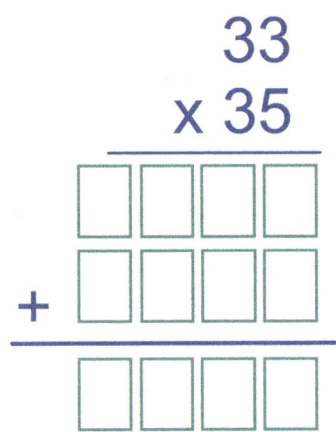

Solve equations below.

12. $5\overline{)25}$

   5 x ? = 25

13. $\dfrac{35}{5} =$ _____

   5 x ? = 35

14.   36 ÷ 6 = _____

   6 x ? = 36

Lesson 33

More Word Problems

Since word problems are so important in math, we're going to look more at these type of problems.

Remember, you want to find and underline the important information and create an equation that you can solve.

(33.1) There are seven days in a week. How many days are in eight weeks and three days?

(33.2) If 42 days passed by, how many weeks passed by?

(33.3) January has 31 days, February has 28 days, and March has 31 days. How many days are in the first three months of the year?

Solve the equations below.

(33.4) $\dfrac{44}{11} =$ ___

(33.5) $\dfrac{100}{10} =$ ___

(33.6) $\dfrac{21}{7} =$ ___

Just the Facts! Workbook

Lesson 33: More Word Problems

(33.7)   Solve the following two ways.
1) Solve using the **distributive property**.
2) Solve by first doing the addition inside the parenthesis.

$$3(4 + 2)$$

3(4 + 2)   3(4 + 2)

____ + ____   4 + 2 = ____

= ____   3 x ____ = ____

Write the **standard** number by un-expanding the numbers.

(33.8)   30,000 + 400 + 20 + 5 = __ __ , __ __ __

(33.9)   50,000 + 100 + 40 = __ __ , __ __ __

(33.10) 600,000 + 1000 + 300 = __ __ __ , __ __ __

Solve the division problems below.

(33.11)   (33.12)   (33.13)

$\dfrac{35}{7}$ = ___   6 ⟌ 60   72 ÷ 9 = ___

___ x ? = ___   ___ x ? = ___   ___ x ? = ___

Lesson 33: More Word Problems

Independent Work - Page 1

1. Sam went to sleep away camp for 21 days. How many weeks was Sam at camp?

2. Emily stayed with her grandparents for five weeks and three days. How many days in all did Emily stay with her grandparents?

Solve the division problems below.

3. $9\overline{)81}$

4. $5\overline{)15}$

5. $3\overline{)21}$

6. $\dfrac{20}{4} = $ _____

7. $\dfrac{20}{5} = $ _____

8. $\dfrac{35}{7} = $ _____

Lesson 33: More Word Problems

Independent Work - Page 2

9. $3(9 + 2)$

$3(9 + 2)$ $3(9 + 2)$

___ + ___ $9 + 2 =$ ___

= ___ $3 \times$ ___ = ___

Fill in the blanks to expand the numbers below.

10. $2{,}735 =$ ___000 + ___00 + ___0 + ___

11. $5{,}820 =$ ___000 + ___00 + ___0

Expand the number below.

12. $9{,}617 =$ _____

_____

Circle the equation that is equal to the one on the left.

13. $4^3 =$ $4 \times 4 \times 4$ $4 + 4 + 4$ 4×4

14. $3^2 =$ $2 \times 3 \times 2$ $3 + 3$ 3×3

Lesson 34 — Fractions

A **fraction** is a smaller, equal part of a whole something. The something can be many things, such as a shape or an object.

When we write a fraction it is written just like a division problem, with one number over another number.

In the circle below, one piece, out of 2, is shaded.

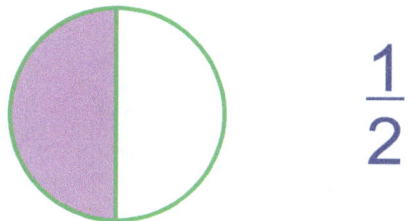

$\dfrac{1}{2}$ ← This line means " out of ".

The 1 over 2 above is a **fraction**.

A **fraction** is not a whole number, like 1, 2, and 3. If you divide the 1 by the 2, you will get a decimal, 0.5.

Note that decimals are not covered in this book, other than their mention here.

What fraction of the rectangle below is shaded?

One out of four smaller, *equal* rectangles inside the larger rectangle is shaded. So, the answer is 1/4.

(34.1)   What fraction of the rectangle is **not** shaded? _____

Lesson 34: Fractions

Below is a whole square that is divided into four equal pieces.

(34.2) Shade 1/4 of the square.

(34.3) Shade 4/4 of the square.

Shade the rectangles below according to the given fraction.

(34.4) 1/3   (34.5) 2/3   (34.6) 3/3

Write the fraction for the shaded area of the rectangles below.

(34.7)   (34.8)   (34.9)

   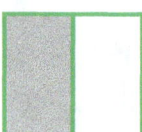   

_____   _____   _____

Just the Facts! Workbook

Lesson 34: Fractions

(34.10)

A pizza is sliced into eight pieces, and Jim ate three pieces. What fraction of the pizza did Jim eat?

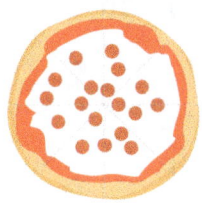 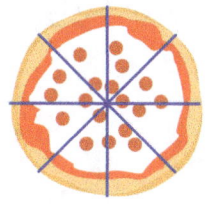 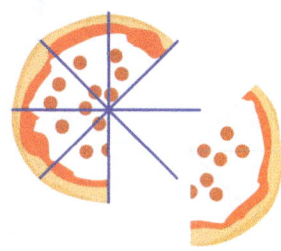

(34.11)

Divide the square into four equal parts.

Now shade 3/4 of the square.

(34.12)

Remember, the pieces must be divide **equally.**

Is 1/4 of the triangle below shaded?

Just the Facts! Workbook

Lesson 34: Fractions

Independent Work - Page 1

1. Jim ate 1/4 of a pizza. Divide the pizza up into 4 pieces and shade the amount that Jim ate.

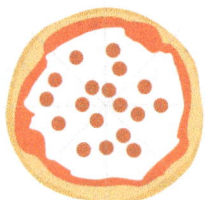

2. Write a fraction to show how many apples fell off the tree.

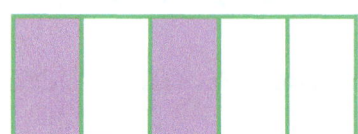

3. Write a fraction to show how many rectangles are shaded.

4. Write a fraction to show how many rectangles are **not** shaded.

Lesson 34: Fractions

Independent Work - Page 2

Solve the problems below.

5. 9)81 ⇒ 9 x ? = 81   [ ]

6. 3)21 ⇒ 3 x ? = 21   [ ]

7. 5)35 ⇒ 5 x ? = 35   [ ]

8. 15 ÷ 5 = ⇒ 5 x ? = 15   [ ]

9. 36 ÷ 6 = ⇒ 6 x ? = 36   [ ]

10. $\frac{20}{2}$ = ⇒ 2 x ? = 20   [ ]

11. $\frac{42}{7}$ = ⇒ 7 x ? = 42   [ ]

12. $\frac{56}{8}$ = ⇒ 8 x ? = 56   [ ]

Just the Facts! Workbook

Lesson 35: Fractions on a Number Line & Ruler

We can apply fractions to number lines.
Below we use a fraction to describe a piece of a rectangle.

On a number line, if we divide one whole segment into four equal parts, we can plot 1/4.

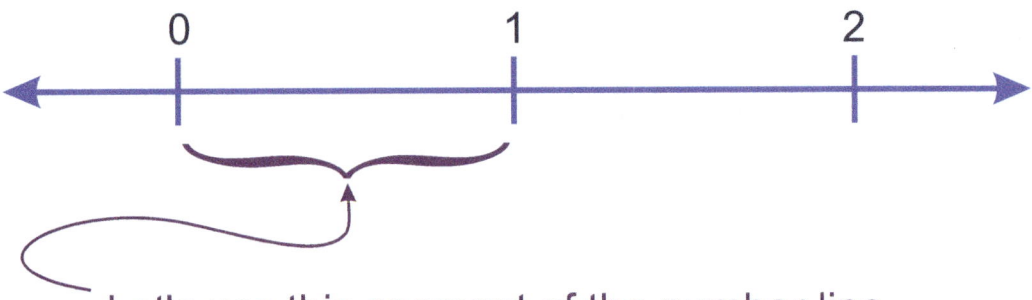

Let's use this segment of the number line.

Below, the segment is divided into four equally sized parts.

We can plot the point at the end of the first 1/4 spot in the segment.

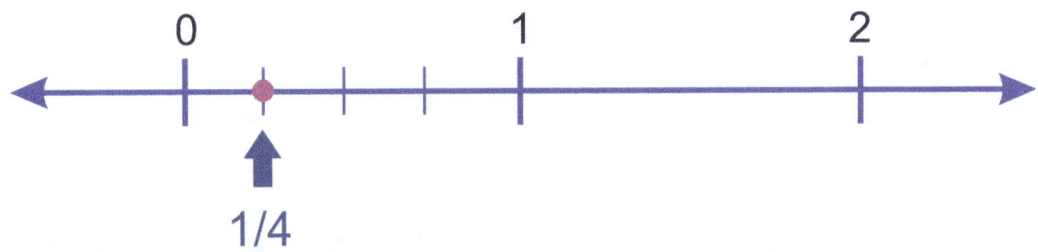

Just the Facts! Workbook

Lesson 35: Fractions on a Number Line & Ruler

(35.1) **Fill in the blanks with the fractions for the spot indicated by the arrows.**

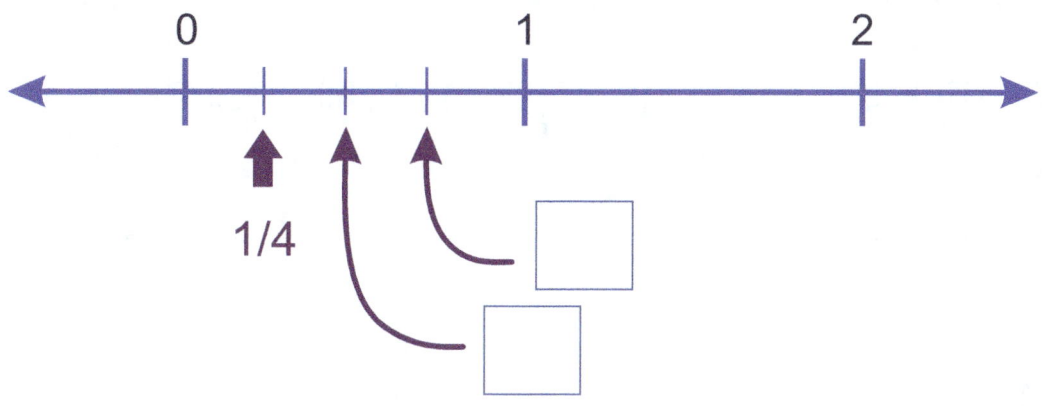

Mixing Fractions with Whole Numbers

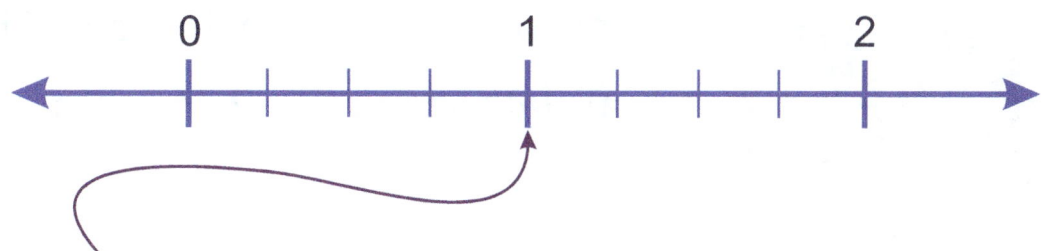

This spot is 4/4, and four divided by four is one.
4 ÷ 4 = 1, because four goes into four one time.

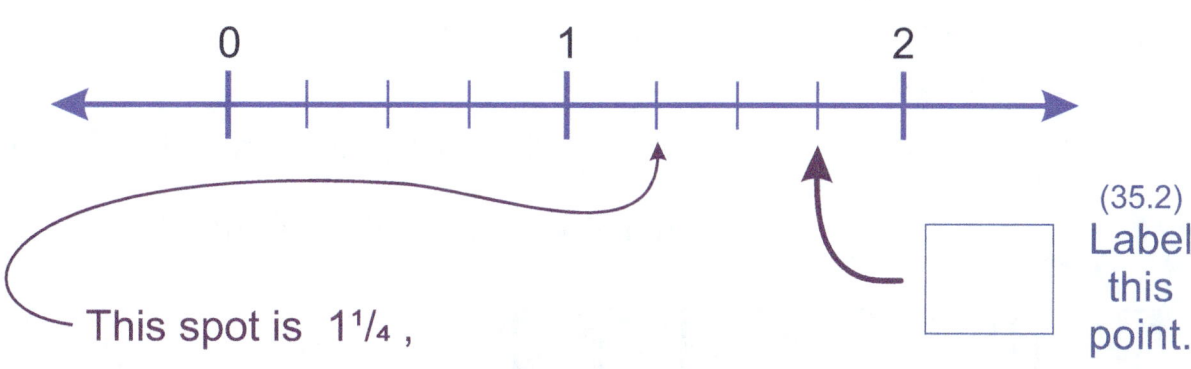

This spot is 1¼, which is read as "one and one fourth".

(35.2) Label this point.

Lesson 35: Fractions on a Number Line & Ruler

Below is an image of a ruler, that has been made bigger so that you can see all the lines clearly. Thus, this ruler is not to scale.

A ruler is just like a number line. Where each inch is equally divided into sixteen parts, four parts and two parts.

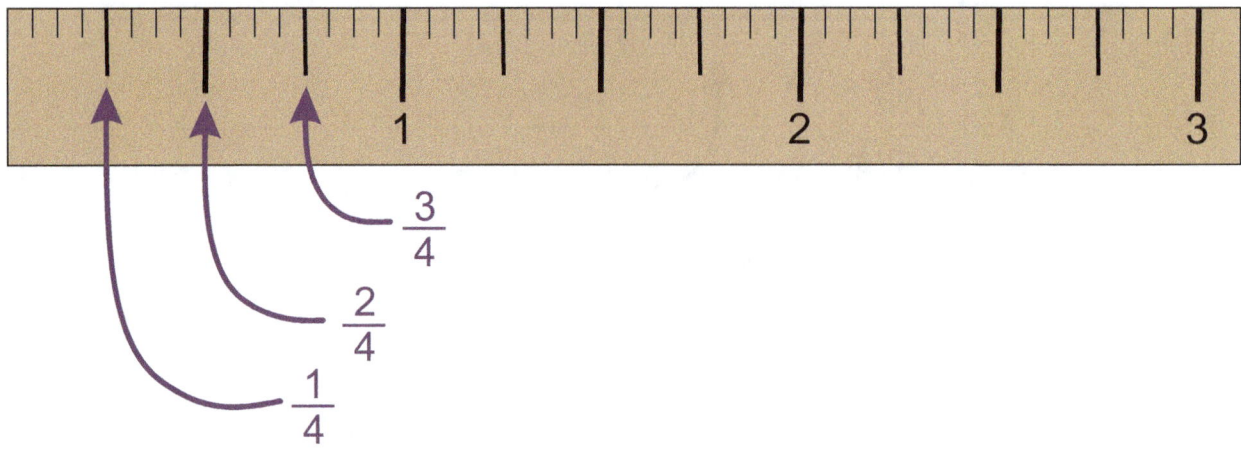

(35.3) What is the length of the lady bug?

(a) 1/4 inch (b) 1/2 inch (c) 3/4 inch

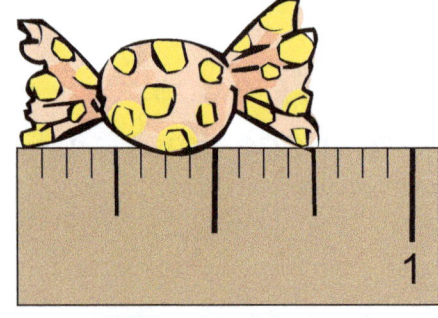

(35.4) What is the length of the candy?

(a) 1/4 inch (b) 3/4 inch (c) 1¼ inch

(35.5) What is the length of the screw?

(a) 1/4 inch (b) 3/4 inch (c) 1¼ inch

Just the Facts! Workbook

Lesson 35: Fractions on a Number Line & Ruler

Independent Work - Page 1

Circle true or false for the questions below.

1. True or False: 3/4 of the triangle is shaded.

2.  True or False: 3/10 of the apples are in the tree.

3. True or False: 3/10 of the squares are shaded.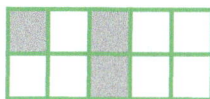

4.  True or False: 1/2 of the triangle is shaded.

5. True or False: 5/7 of the coins are dimes.

6.  True or False: 4/8 of the pizza pie has pepperoni.

7. True or False: 3/5 of the circle is shaded.

Lesson 35: Fractions on a Number Line & Ruler

Independent Work - Page 2

Circle the number for the spot on the number line indicated.

8.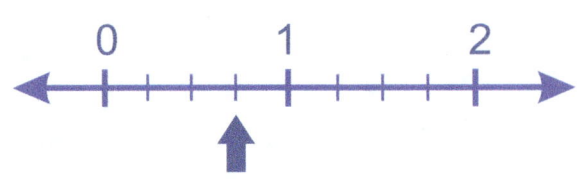

(a) 1/4   (b) 3/4   (c) 1 ¼   (d) 1 ¾

9.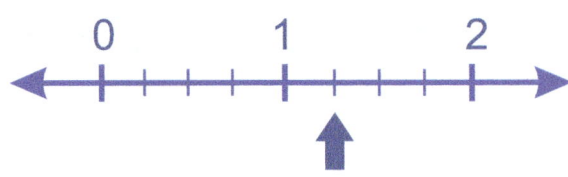

(a) 1/4   (b) 3/4   (c) 1 ¼   (d) 1 ¾

10.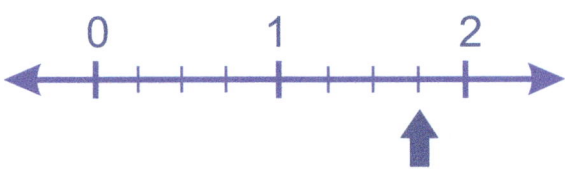

(a) 1/4   (b) 3/4   (c) 1 ¼   (d) 1 ¾

11. What is the length of the fish?

(a) 1/4 in. (b) 3/4 in. (c) 1 ¼ in.

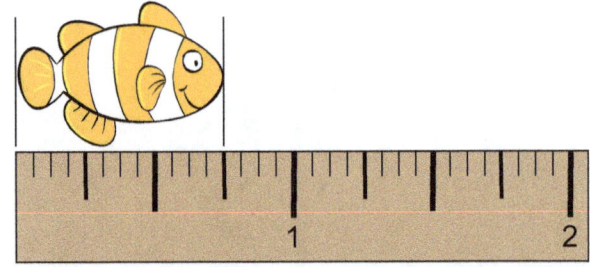

12. What is the length of the nail?

(a) 1/4 in. (b) 3/4 in. (c) 1 ¼ in.

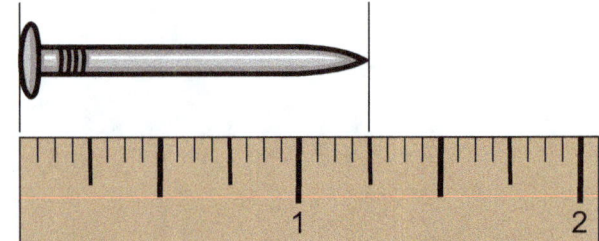

Just the Facts! Workbook

Lesson 36: Comparing Fractions

In this lesson we will compare fractions.

Below, both rectangles are the same size, and both are divided equally into three pieces.

Which rectangle has more area shaded?

$\dfrac{2}{3}$ of the rectangle > $\dfrac{1}{3}$ of the rectangle

Since both rectangles are the same size, we can *see* that 2/3 is larger than 1/3.

Complete the equations with >, < or = to make the statements true.

(36.1) $\dfrac{3}{10} \bigcirc \dfrac{4}{10}$

(36.2) $\dfrac{7}{10} \bigcirc \dfrac{5}{10}$

(36.3) $\dfrac{9}{10} \bigcirc \dfrac{5}{10}$

(36.4) $\dfrac{5}{10} \bigcirc \dfrac{3}{10}$

Just the Facts! Workbook

Lesson 36: Comparing Fractions

The parts of fractions have special names. The top number is called the **numerator,** and the bottom number is called the **denominator.**

$$\frac{7}{16}$$ ← numerator
← denominator - think 'd' for "downstairs"

In the exercises on the previous page, you can see that if the denominators are the same, you can easily compare the fractions by looking only at the numerators.

What if the denominators are different?

Complete the equations with >, < or = to make the statements true.

(36.5)

$$\frac{7}{10} \bigcirc \frac{1}{4}$$

(36.6)

$$\frac{4}{6} \bigcirc \frac{3}{10}$$

(36.7)

$$\frac{5}{10} \bigcirc \frac{4}{8}$$

(36.8)

$$\frac{4}{8} \bigcirc \frac{3}{6}$$

Notice how fractions can be **equal** to other fractions.

Just the Facts! Workbook

Lesson 36: Comparing Fractions

Independent Work - Page 1

Circle the name for the red, bold number.

1.	$\frac{\mathbf{10}}{5} = 2$	numerator	denominator	quotient
2.	$5 \times 2 = \mathbf{10}$	product	factor	sum
3.	$\frac{10}{\mathbf{5}} = 2$	denominator	quotient	numerator
4.	$\frac{10}{5} = \mathbf{2}$	product	sum	quotient
5.	$5 \times \mathbf{2} = 10$	factor	product	sum

Complete the equations with >, < or = to make the statements true.

6. $\frac{6}{10} \bigcirc \frac{2}{10}$

7. $\frac{1}{10} \bigcirc \frac{4}{10}$

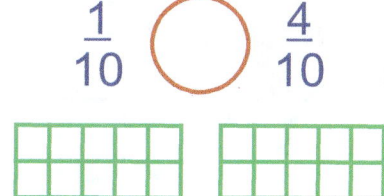

8. $\frac{4}{8} \bigcirc \frac{3}{4}$

9. $\frac{4}{5} \bigcirc \frac{7}{10}$

Lesson 36: Comparing Fractions

Independent Work - Page 2

10. Jim ate 1/4 of a pizza, and Sue ate 3/8 of a pizza. Who ate more?

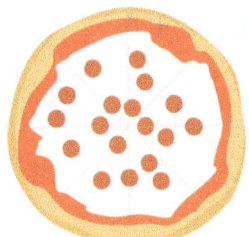 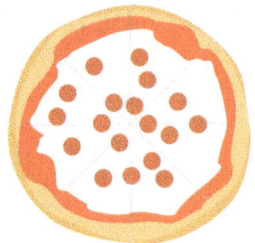

11. Write a fraction for the amount of apples in the tree.

12. How long is the green line?

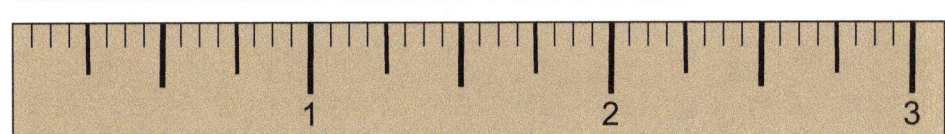

Solve the equations below.

13. $\dfrac{36}{6} =$ ___

14. $\dfrac{42}{7} =$ ___

15. $\dfrac{48}{8} =$ ___

Just the Facts! Workbook

Lesson 37: Telling Time on an Analog Clock

There are two types of clocks, **analog** and **digital**. Digital clocks will tell you the time in numbers only, whereas an analog clock you have to use the hands on the clock to determine the time.

Analog Clock

Digital Clock

Telling Time on an Analog Clock

9 o'clock or 9:00

On an analog clock there are two hands, a long one and a short one.

The short one points to the hours, and the long one points to the minutes.

When the long hand (the minute hand) is pointing to the 12, then the short hand will be directly pointing to a number. That number tells you what hour it is. The time on this clock is 9 o'clock.

What time is it?

(37.1) 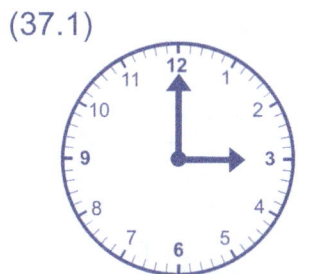 ____:____

(37.2) 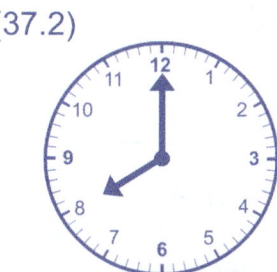 ____:____

(37.3) 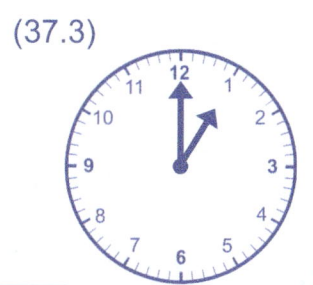 ____:____

Just the Facts! Workbook

Lesson 37: Telling Time on an Analog Clock

Both hands go around the clock as shown below.
This is called **clockwise**.

8:20

As the **minute** hand (the long hand) goes around the clock, the **hour** hand (the short hand) slowly goes to the next number. That means that the hour hand may not point directly to a number, it may be *between* numbers.

<u>Step 1</u>: Read the number that the hour hand (the short hand) was at (the hour it would have been if the minute had was pointing to the 12). In the clock above it would be the number 8.

<u>Step 2</u>: Starting at the 12, count by fives and then, if needed, count by ones, to where the minute hand (the long hand) is pointing to. In the clock above it would be 20, thus the time is 8:20.

What time is it?

(37.4)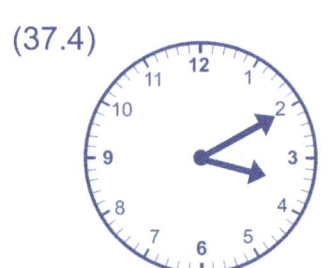

The hour: _____

The minute: _____

The time: ___:___

(37.5)

The hour: _____

The minute: _____

The time: ___:___

(37.6)

The hour: _____

The minute: _____

The time: ___:___

Just the Facts! Workbook

Lesson 37: Telling Time on an Analog Clock

Telling Time with Fractions

When the minute hand (the long hand) is pointing to a 3, 6, or 9, we often use the word for specific fractions to say the time.

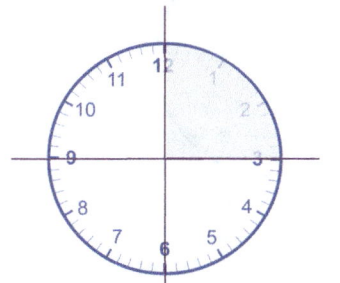 $\frac{1}{4}$ = quarter   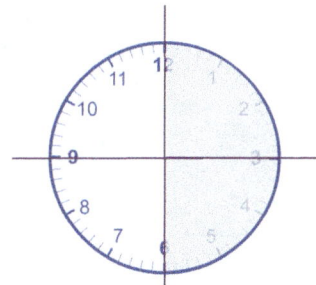 $\frac{1}{2}$ = half

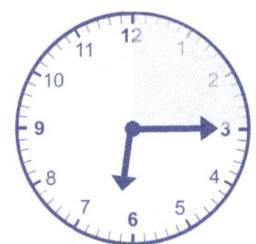      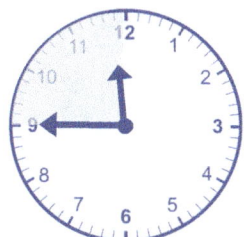

6:15 or quarter past six   8:30 or half past eight   11:45 or quarter to twelve

Write the time the standard way, and then circle the correct time using fractions as words.

(37.7)   (37.8)   (37.9)

   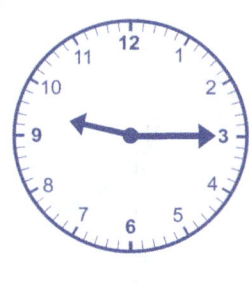   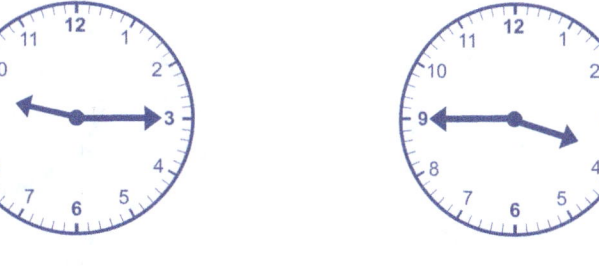

____:____   ____:____   ____:____

(a) half past two   (a) quarter past ten   (a) quarter to four

(b) half past one   (b) quarter past nine   (b) half past three

Just the Facts! Workbook

Lesson 37: Telling Time on an Analog Clock

Independent Work - Page 1

What time is it?

1. [clock image]

2.

3.

___:___   ___:___   ___:___

Circle the time indicated for each clock below.

4.

(a) 1:05

(b) 7:05

(c) 8:05

5.

(a) 12:20

(b) 4:58

(c) 11:20

6.

(a) 6:35

(b) 6:30

(c) 7:32

7.

(a) 4:50

(b) 5:45

(c) 4:48

8.

(a) 12:20

(b) 4:58

(c) 11:20

9.

(a) 6:35

(b) 6:30

(c) 7:32

Lesson 37: Telling Time on an Analog Clock

Independent Work - Page 2

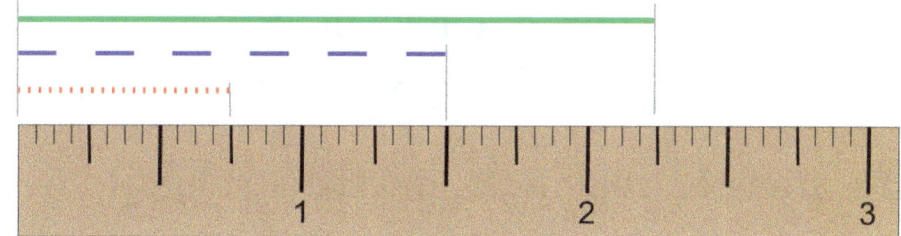

10. How long is the solid line? _____ inch

11. How long is the dashed line? _____ inch

12. How long is the dotted line? _____ inch

---

13. Is this true? $\dfrac{1}{2} = \dfrac{2}{4}$ (a) yes   (b) no

14.
$$\begin{array}{r} 67 \\ \times\ 29 \\ \hline \end{array}$$

15. $4\overline{)32}$

16. $\dfrac{40}{8} = \underline{}$

17. $56 \div 8 = \underline{}$

18. $64 \div 8 = \underline{}$

19. $5\overline{)45}$

20. $\dfrac{18}{3} = \underline{}$

Just the Facts! Workbook

Lesson 38 — Area of a Square & Rectangle

Area is a measurement of the surface of
a flat (two dimensional) shape.

To find the **area** of a **rectangle,** multiply the **length** by the **width.**

length

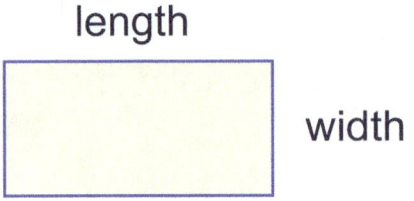

width

Area = (length)(width)

(38.1) How many units are in the rectangle below?

5 units

4 units   (5)(4) = _____ units

We can check to see if our answer is correct.
Below is a rectangle that is divided into the
units as described in the rectangle above.

(38.2) Count the small rectangles inside the large rectangle.

5 units

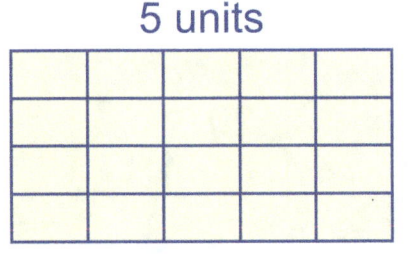

4 units

There are _____ units.

Just the Facts! Workbook   - 204 -

Lesson 38: Area of a Square & Rectangle

The units for area are squared, much like when an exponent is 2. 4^2 can be read as four to the second power or four **squared**.

Area = (5 units)(4 units) = 20 square units

5 units

4 units

(38.3) Mary's bedroom is 10 feet long and 9 feet wide. She wants to buy a new carpet for her room. What is the area of the carpet that she will need?

10 feet

9 feet

_____ x _____ = _____

Area = _____ square feet

The **area of a square** is just like a rectangle. However, since a square has four *equal* sides, you only need to know the length of **one** side to find the area.

(38.4) What is the **area** of the square below?

5 inches

_____ x _____ = _____

Area = _____ square inches

Just the Facts! Workbook

Lesson 38: Area of a Square & Rectangle

(38.5) Use the symbols < , > or = to make the statement true.

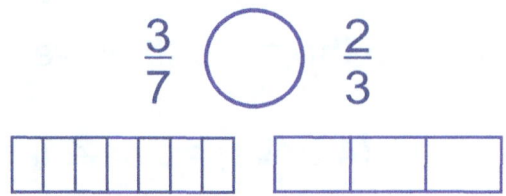

(38.6) If the **denominators** are **equal**, would you need to draw it out as in (38.5) above? (a) yes (b) no

(38.7) Use the symbols < , > or = to make the statement true.

$$\frac{3}{7} \bigcirc \frac{2}{7}$$

What time is it?

(38.8)

____:____

(38.9)

____:____

(38.10)

____:____

(38.11) Draw in the hands for half past nine.

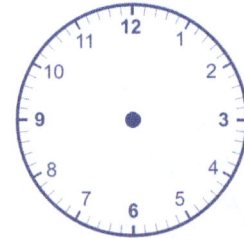

(38.12) Draw in the hands for quarter to four.

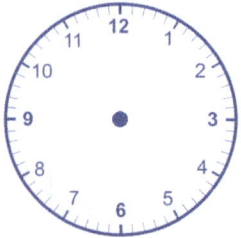

Lesson 38: Area of a Square & Rectangle

Independent Work - Page 1

Find the areas for the rectangles below.

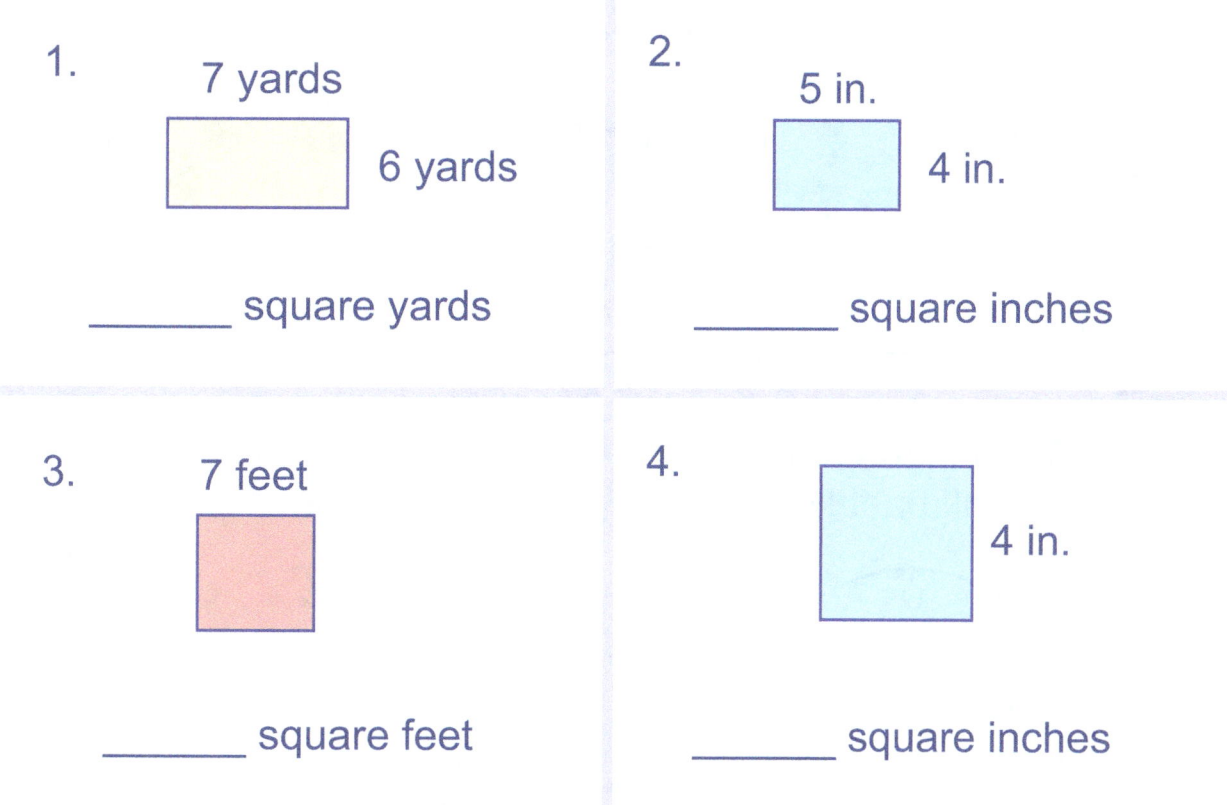

1. 7 yards / 6 yards
   _____ square yards

2. 5 in. / 4 in.
   _____ square inches

3. 7 feet
   _____ square feet

4. 4 in.
   _____ square inches

5. Mrs. Henley's yard is <u>100 feet</u> long by <u>65 feet</u> wide. She needs to buy grass seed, but each bag is for 1000 square feet. How many bags does Mrs. Henley need to purchase?

Mrs. Henley's yard.

Lesson 38: Area of a Square & Rectangle

Independent Work - Page 2

What time is it?

6.

____:____

7.

____:____

8.

____:____

9. Draw in the hands for half past two.
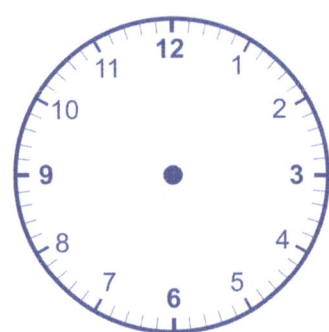

10. Draw in the hands for quarter to six.
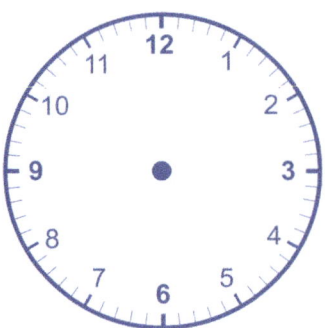

Use the symbols >, <, or = to make the statements below true.

11. $\frac{4}{7}$ ◯ $\frac{2}{7}$

12. 9*7 ◯ 8*4

13. $\frac{8}{9}$ ◯ $\frac{4}{9}$

14. $\frac{80}{10}$ ◯ 2*4

15. 1 ¼ ◯ 1 ¾

16. $\frac{20}{4}$ ◯ $\frac{25}{5}$

Just the Facts! Workbook

Lesson 39

Perimeter of a Closed Shape

Perimeter is the measurement around a closed shape.

To find the **perimeter** any shape, simply measure the line that goes all the way around the outer edge of the shape.

For a rectangle:

length

width   width

length

Perimeter of a rectangle = length + length + width + width, or
Perimeter of a rectangle = 2(length) + 2(width)

(39.1) What is the **perimeter** (P) of the rectangle below?

5 units

4 units

P = 5 + 5 + 4 + 4, or
P = 2(5) + 2(4)

The perimeter is

_____ units.

(39.2) What is the **perimeter** (P) of the square below?

10 feet

P = 10 + 10 + 10 + 10, or
P = 2(10) + 2(10), or 4(10)

The perimeter is

_____ feet.

Just the Facts! Workbook - 209 -

Lesson 39: Perimeter of a Square & Rectangle

(39.3) What is the **perimeter** (P) of the triangle below?

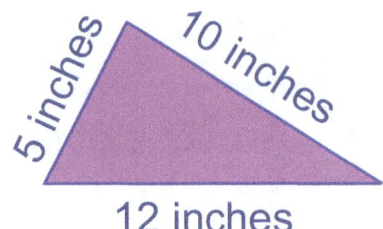

The perimeter is

_____ inches.

(39.4) What is the **perimeter** (P) of the shape below?

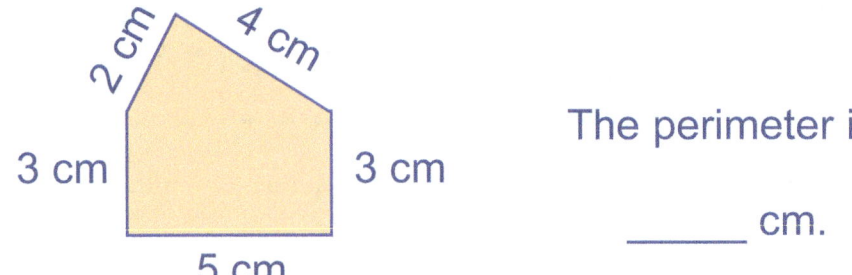

The perimeter is

_____ cm.

(39.5) Zach's yard is 200 feet long and 90 feet wide. He wants to buy a fence that goes around the entire yard. How many feet of fence does Zach need to buy?

Zach's yard

Zach needs _____ feet of fence.

Lesson 39: Perimeter of a Square & Rectangle

(39.6) What time is it?

____:____

(39.7) What time is it?

____:____

(39.8) What time is it?

____:____

Use the symbols <, >, or = to make the statements below true.

(39.9) $\frac{4}{9}$ ◯ $\frac{3}{5}$

$\frac{4}{9}$ ▯▯▯▯▯▯▯▯▯
$\frac{3}{5}$ ▯▯▯▯▯

(39.10) $\frac{4}{8}$ ◯ $\frac{3}{6}$

$\frac{4}{8}$ ▯▯▯▯▯▯▯▯
$\frac{3}{6}$ ▯▯▯▯▯▯

(39.11)

$$876 \times 359$$

In the answer, what number is in the **ten thousands** spot?

____

Just the Facts! Workbook

Lesson 39: Perimeter of a Square & Rectangle

Independent Work - Page 1

Find the **perimeter** and the **area** of the shapes below.

1.
1 in. [rectangle] 10 in.

A = length x width
P = 2(length)+2(width)

(a) Perimeter = _____ inches

(b) Area = _____ square inches

2.
7 yards
[rectangle] 6 yards

A = length x width
P = 2(length)+2(width)

(a) Perimeter = _____ yards

(b) Area = _____ square yards

Find the **perimeter** of the shapes below.

3.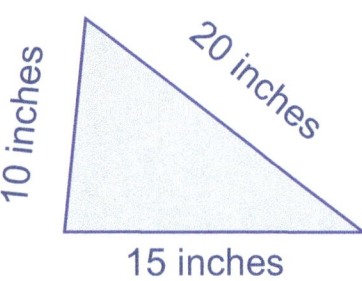

4.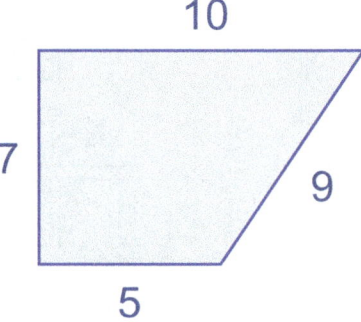

P = _____ inches

P = _____ units

Just the Facts! Workbook

Lesson 39: Perimeter of a Square & Rectangle

Independent Work - Page 2

5. What time is it?   6. What time is it?   7. What time is it?

____:____   ____:____   ____:____

Use the symbols < , >, or = to make the statements below true.

8. $\frac{4}{5}$ ◯ $\frac{4}{6}$ 9. $\frac{1}{2}$ ◯ $\frac{10}{20}$

10.

$$\begin{array}{r} 698 \\ \times\ 87 \\ \hline \end{array}$$

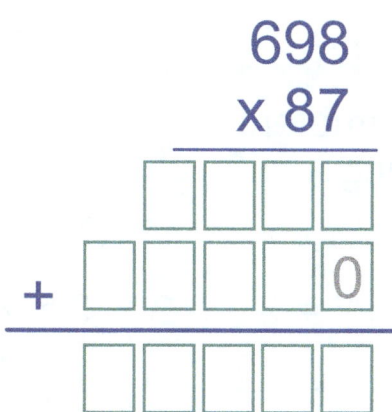

In the answer, what number is in the **thousands** spot?

_____

Lesson 40: Finding a Length, Given the Area

If we know the **area**, we can find a missing value for the length by using multiplication and division.

(40.1) Below is a square with an area of **25 square inches**. What are the lengths of the sides?

? in.

? × ? = 25
Remember, this is a **square,** so all sides are **equal.**

Each side has a length of

_____ inches.

(40.2) What is the length of the unknown side of the rectangle if the area is **30 square feet**?

5 feet

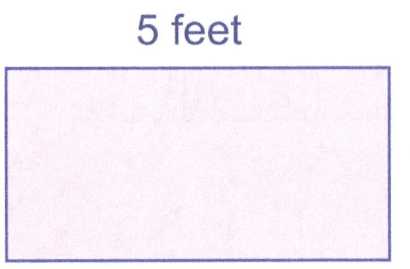

? feet

5 × ? = 30   or   30 ÷ 5 = ?

The unknown side has a

length of _____ feet.

(40.3) Area = 3 square centimeters
How long is the missing side?

3 cm

 ? cm

3 × ? = 3   or   3 ÷ 3 = ?

The unknown side has a length of _____ centimeters.

Just the Facts! Workbook

Lesson 40: Finding a Length, Given the Area

(40.4)

Jim wants to make a garden, but first he needs to put down good soil. He has one bag of soil that covers an area of 50 square feet. Does Jim have enough soil for his garden?

10 feet

Jim's Garden   6 feet

(40.5)

What is the **perimeter (P)** and **area (A)** of the shape below?

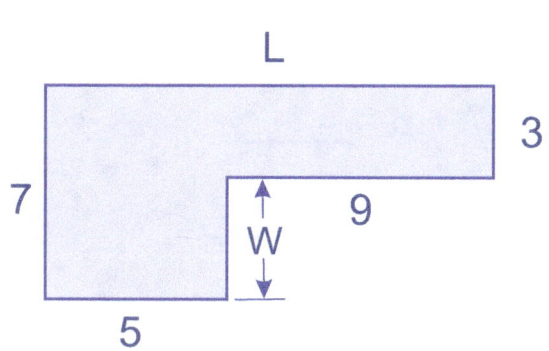

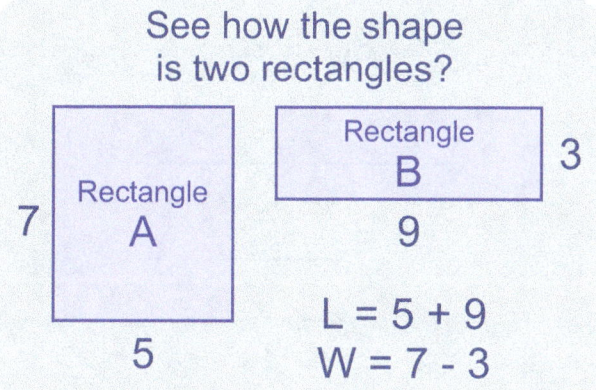

See how the shape is two rectangles?

$L = 5 + 9$
$W = 7 - 3$

To find P, find L and W then add all the sides.
To find A, find the area of each smaller rectangle, and add them.

(a) Perimeter = _____ units   (b) Area = _____ sq. units

(40.6) Round 9,166 to the nearest 10. _____

(40.7) Round 3,482 to the nearest 100. _____

(40.8) Round 11,452 to the nearest 100. _____

Lesson 40: Finding a Length, Given the Area

(40.9)	(40.10)	(40.11)	(40.12)	(40.13)
500 x 4	60 x 30	700 x 9	90 x 3	800 x 7

(40.14)   Solve the problem using the distributive property and by adding and then multiplying.

$$8(7 + 6)$$

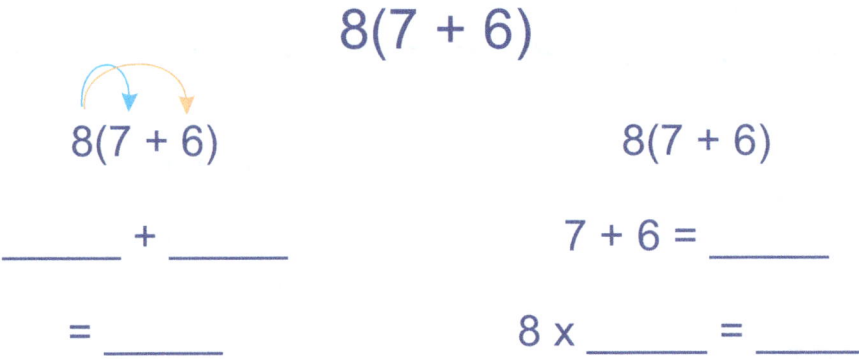

(40.15) What time is it?

____ : ____

(40.16) What fraction of fruit is bananas?

(40.17) What fraction of fruit is apples?

Lesson 40: Finding a Length, Given the Area

Independent Work - Page 1

1. The square below has an area of 64 square feet. What are the lengths of all the sides?

? feet

A = 64 sq. ft.

Area = Length x Width

64 = _____ x _____

The sides = _____ feet

2. The square below has an area of 100 square feet. What are the lengths of all the sides?

? feet

A = 100 sq. ft.

Area = Length x Width

100 = _____ x _____

The sides = _____ feet

3. Emma wants to paint a wall in her room red. The paint can instructions says that it will cover **40 square feet**. Emma's wall is <u>8 feet</u> high by <u>11 feet</u> long.

11 ft

8 ft

What is the area to paint?

_____ square feet

If Emma has only one can of paint, does she have enough?

yes   no

Just the Facts! Workbook

Lesson 40: Finding a Length, Given the Area

Independent Work - Page 2

4.	5.	6.	7.	8.
300 x 8	80 x 90	800 x 4	900 x 40	60 x 7

9. Solve the problem using the distributive property and by adding and then multiplying.

$$4(8 + 9)$$

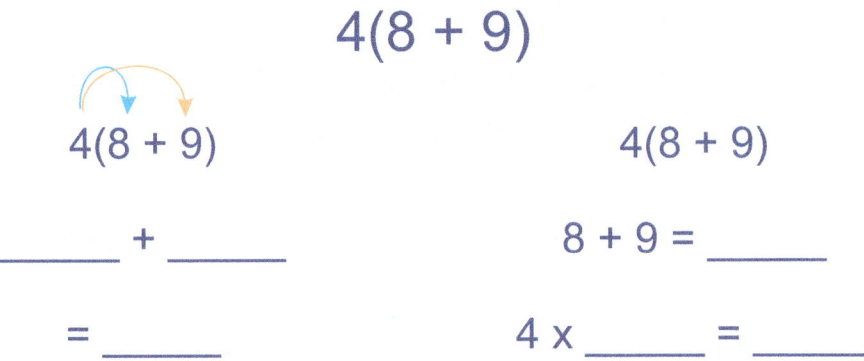

4(8 + 9)   4(8 + 9)

____ + ____   8 + 9 = ____

= ____   4 x ____ = ____

10. What time is it?

____:____

11. What fraction of fruit is bananas?

12. What fraction of fruit is apples?

Lesson 41: Volume

Volume is the amount of *space* that something takes.

For now, we will only look at the volume of **rectangular prisms** and **cubes**.

A **rectangular prism** is like a rectangle and a **cube** is like a square, however they are *three* dimensional instead of two dimensional.

cube

rectangular prism

All side lengths are equal.

All side lengths are *not* equal.

Volume (V) of a Rectangular Prism

V = height x length x width

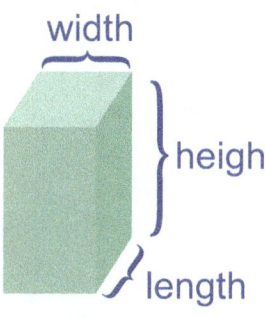

4 cm

8 cm

3 cm

V = 8 cm x 3 cm x 4 cm

V = 24 cm² x 4 cm

V = 96 cm³ or 96 cubic centimeters

Volume is measured in **cubic** units.

Lesson 41: Volume

Volume (V) of a Cube

Since all side lengths on a cube are equal, we don't need to assign a width, length or height. We only need to multiply a side three times.

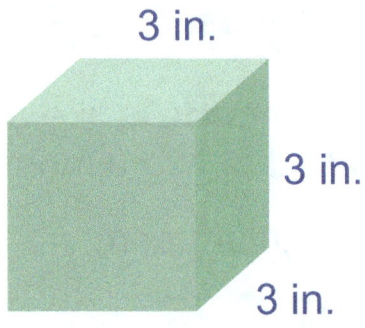

3 in.

3 in.

3 in.

Volume =

3 in. x 3 in. x 3 in.

9 in.² or 9 square inches

9 in.² x 3 in. =
27 in.³ or 27 cubic inches

Notice how we can use exponents for the units.
Also, it is helpful to know that:
1. an exponent of 2 is called "squared",
2. an exponent of 3 is called "cubed".

What is the volume of the cubes below?

(41.1)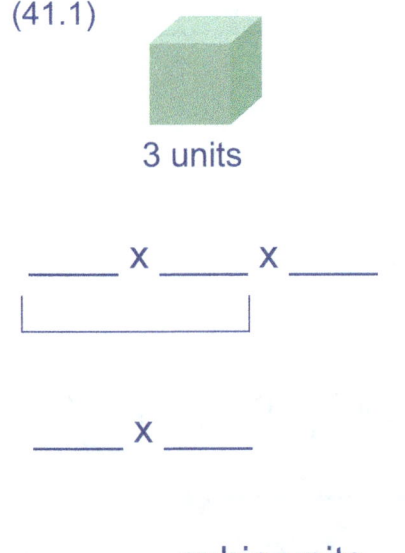
3 units

____ x ____ x ____

____ x ____

_____ cubic units

(41.2)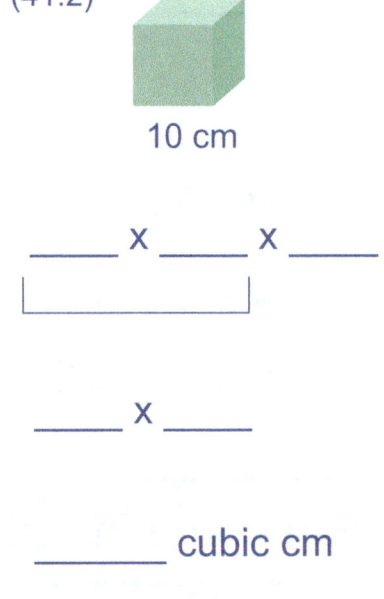
10 cm

____ x ____ x ____

____ x ____

_____ cubic cm

(41.3)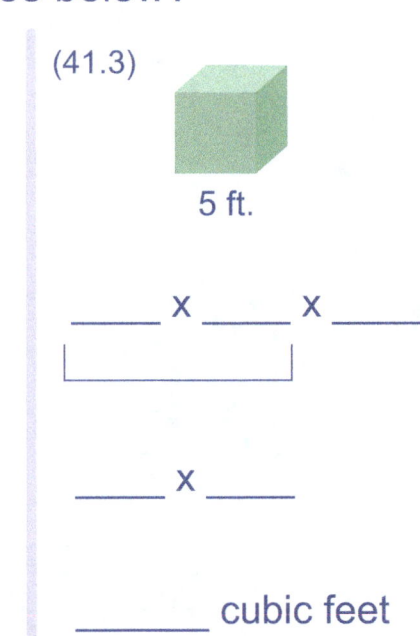
5 ft.

____ x ____ x ____

____ x ____

_____ cubic feet

Lesson 41: Volume

What is the **volume** of the rectangular prisms below?

(41.4)

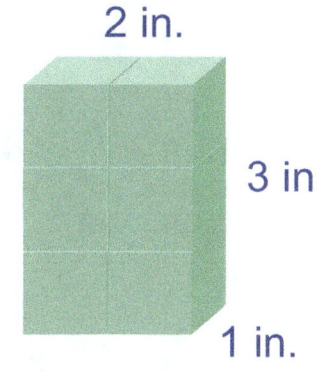

2 in.
3 in.
1 in.

____ x ____ x ____

____ x ____

_____ cubic inches

Count the smaller cubes inside the rectangular prism, how many are there?

_____

(41.5)

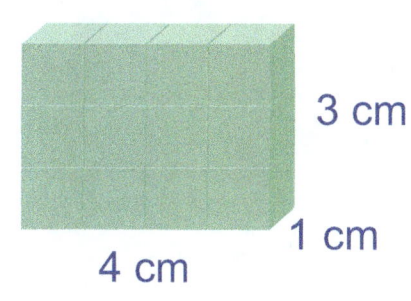

3 cm
1 cm
4 cm

____ x ____ x ____

____ x ____

_____ cubic cm

Count the smaller cubes inside the rectangular prism, how many are there?

_____

(41.6) What is the **volume** of the rectangular prism below?

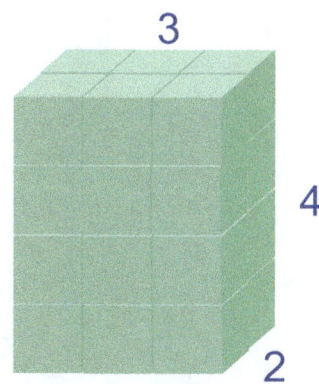

3
4
2

Lesson 41: Volume

Independent Work - Page 1

Find the volume, V, of the cubes below.

1.

5 cm

V = ____ x ____ x ____
|_____|

____ x ____

V = _____ cubic cm

2.

4 in.

V = ____ x ____ x ____
|_____|

____ x ____

V = _____ cubic in.

3. What is the **volume** of the rectangular prism below?

$$V = \text{height} \times \text{length} \times \text{width}$$

2
3
2

V = ____ x ____ x ____
|_____|

____ x ____

V = _____ cubic units

Just the Facts! Workbook

Lesson 41: Volume

Independent Work - Page 2

Use the symbols >, <, or = to make the statements below true.

4.
$\frac{4}{8} \bigcirc \frac{1}{4}$

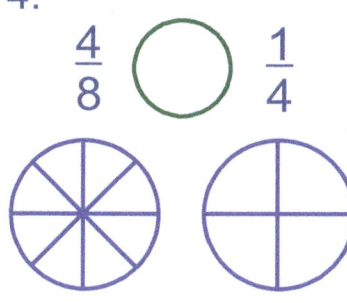

5.
$\frac{4}{8} \bigcirc \frac{2}{4}$

6.
$\frac{2}{8} \bigcirc \frac{1}{4}$

7. Find the perimeter of the shape below.

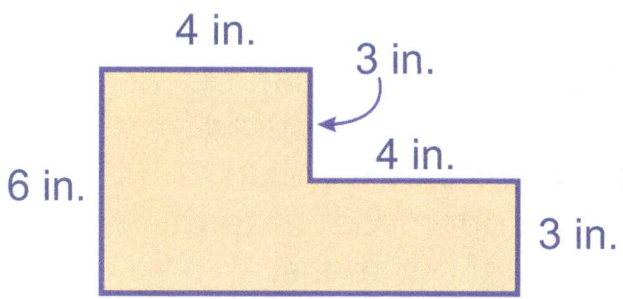

The perimeter is

_____ inches.

8. Find the area of the shape below. Hint: break it up into two rectangles, find each area, and then add the two areas.

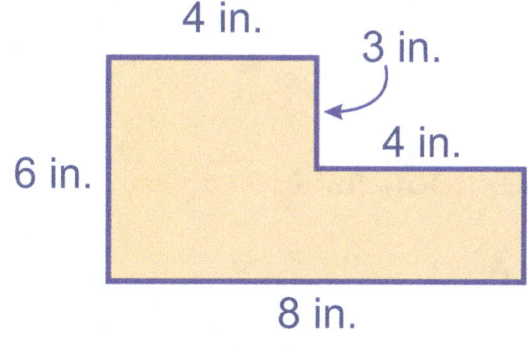

The total area of the shape is

_____ square inches.

Just the Facts! Workbook

Lesson 42: Mass & Weight

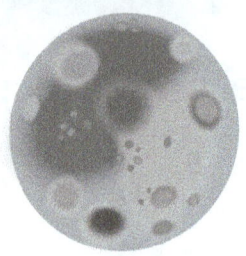

If you weighed 80 pounds on earth, you would weigh 13.23 pounds on the moon. If you live where weight is measured in kilograms, and you weighed 36 kg on earth, you would weigh 5.95 kg on the moon.

Your **mass** would be the same, no matter where you were.

Mass is the amount of matter that an object contains.

Weight is related to mass, depending on the gravity of where the object is.

Here's where it gets confusing. Mass is measured in kilograms (which can be converted to pounds), and weight is measured in Newtons. Here, on earth, when we weigh ourselves, we're really finding our mass, or to be more exact, an *estimate* of our mass.

In the example above, where we compared earth weight to moon weight, we really should be using Newtons instead of kilograms or pounds.

(42.1) Is your **mass** the same on the moon as it is on earth?

(a) yes   (b) no

(42.2) Is your **weight** the same on the moon as it is on earth?

(a) yes   (b) no

Lesson 42: Mass & Weight

(42.3)

If all gold coins have the same mass, circle the picture that will happen if you use a balance scale with three coins on each side.

(a)   (b)   (c)

(42.4)

Jane found 8 gold coins, all having the same mass. She knows that one coin is 9 grams. What is the mass of all 8 coins?

(42.5)

Jack found a bag of 9 iron pellets. When Jack poured the contents of the bag on a scale, he saw the mass was 63 grams. What is the mass of each pellet?

(42.6) The yellow cereal box has 128 grams of cereal, and the red box has 84 grams of cereal. How much more cereal is in the yellow box?

Just the Facts! Workbook

Lesson 42: Mass & Weight

(42.7)   What is the **volume** of the rectangular prism below?

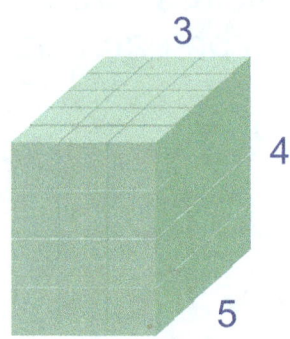

3
4
5

(42.8)   What is the **volume** of the object below?

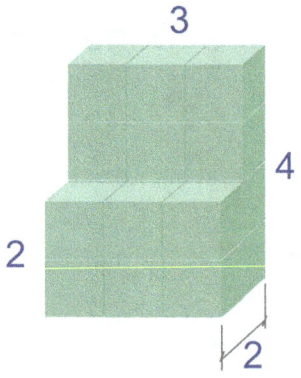

3
4
2
2

(42.9)   What is the length of the unknown side of the rectangle if the area is **99 square feet**?

? feet

9 feet

The unknown side has a length of _____ feet.

(42.10) Find the area (A) and perimeter (P) of the rectangle.

8 cm

5 cm

(a) Area = _____ square cm

(b) Perimeter = _____ cm

Just the Facts! Workbook

Lesson 42: Mass & Weight

Independent Work - Page 1

1. At the store there are two different jars of jam. The shorter jar contains 63 grams of jam, and the taller jar contains 98 grams. How much more jam is in the tall jar?

2. Two sacks are 1 kilogram each. Circle the picture that will happen if you use a balance scale with one sack on each side.

(a)  (b)  (c)

3. Together, four chairs have a combined mass of 100 kg. All of the chairs have the same mass. What is the mass of a single chair?

4. A jar contains 11 lollipops, each having the same mass. If one lollipop has a mass of 8 grams, what is the total mass of the lollipops?

Just the Facts! Workbook

Lesson 42: Mass & Weight

Independent Work - Page 2

5. What is the **volume** of the rectangular prism below?

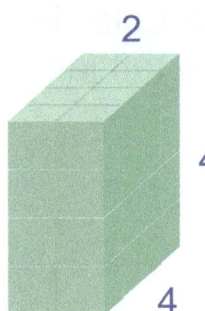

6. What is the **volume** of the object below?

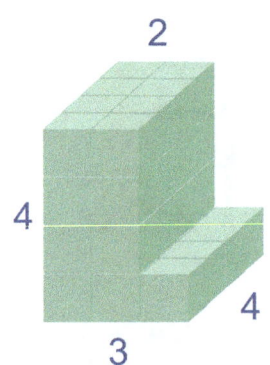

7. The rectangle below has an area of **56 square cm**.

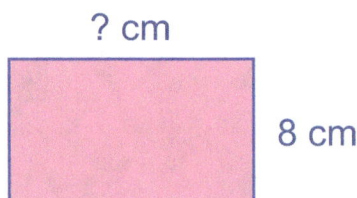

a. What is the length of the unknown side? _____ cm

b. What is the **perimeter** of the rectangle? _____ cm

Lesson 43: Graphing Data

Math is very important for science. Scientists may use math to see what is happening so they can predict what might happen in the future. Scientists often do this by collecting data and using graphs.

There are many types of graphs, but we are going to look at the most popular: **bar graphs** and line **graphs**.

To show the data, **bar graphs** use bars, & **line graphs** use lines.

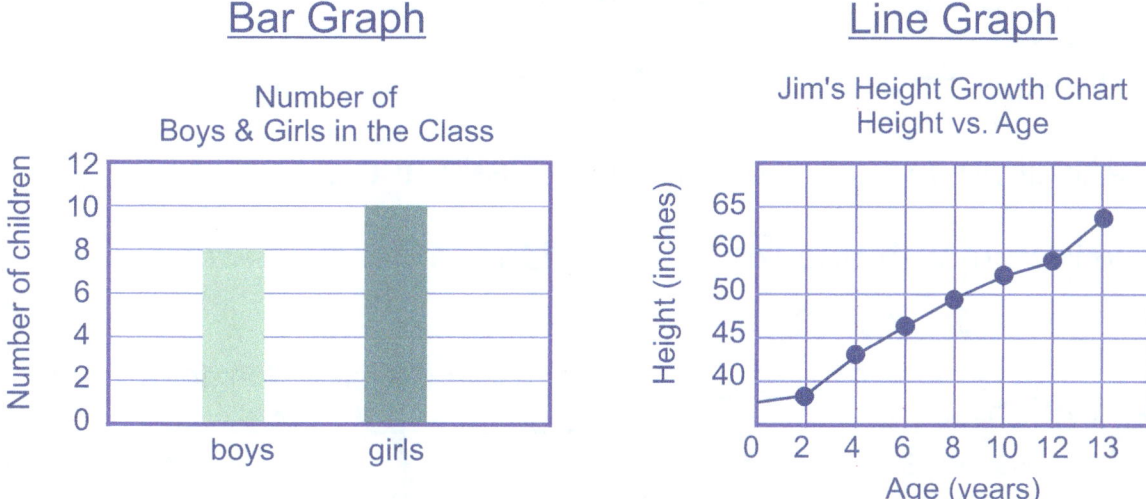

All graphs need to have the information that tells us what is being shown, and what the units are (years, inches, feet, days, etc.). So that when you look at a graph, you can see the information easily.

(43.1) In the bar graph above, are there more boys or girls?

(43.2) Looking at the line graph, do you think Jim will keep getting taller?   (a) yes   (b) no

Just the Facts! Workbook

Lesson 43: Graphing Data

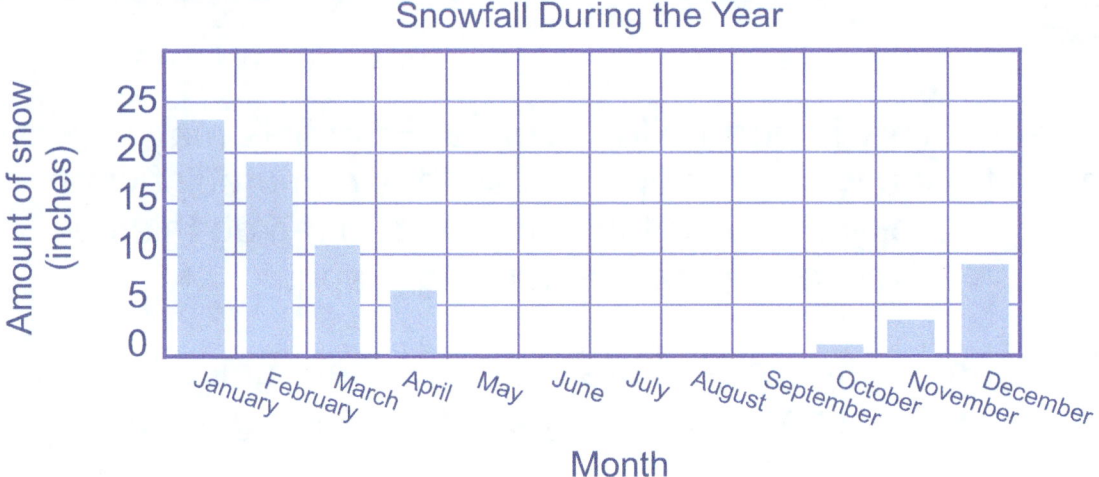

(43.3) Did it snow every month? (a) yes (b) no

(43.4) What month had the most snow?

(43.5) Did April have more than 5 inches of snow? (a) yes (b) no

(43.6) About how much did it snow in December?

(43.7) Which month had more snow, December or March?

(43.8) Which day was the warmest?

(43.9) Which day was the coolest?

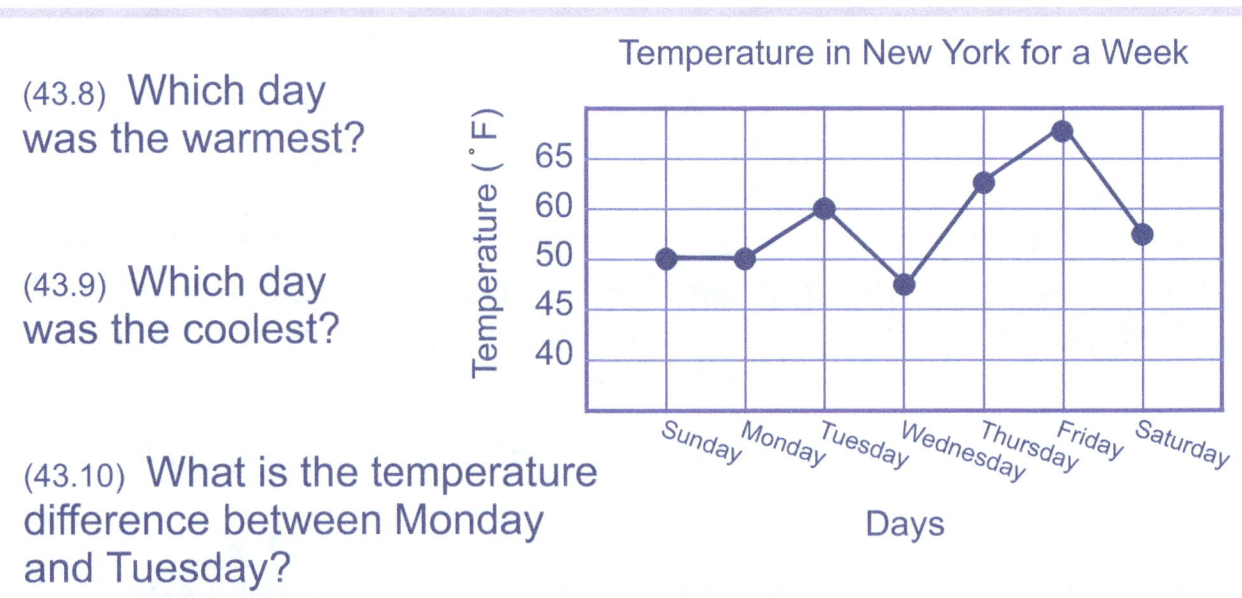

(43.10) What is the temperature difference between Monday and Tuesday?

(43.11) Can you tell how much it rained on Thursday?

Lesson 43: Graphing Data

Scaled Picture Graphs

A scaled picture graph uses pictures, where each picture represents a certain number of something. Below we have a scaled picture graph for holiday cookies.

Holiday Cookies

 = chocolate chip

= ginger bread

= star cookies

Each cookie image represents 5 cookies.

(43.12) How many star cookies were baked?

(43.13) How many more chocolate chip cookies were made than star cookies?

(43.14) How many cookies were baked in all?

(43.15) Draw a bar graph for the following: at the zoo there are 3 elephants, 5 zebras, 4 tigers, and 1 alligator.

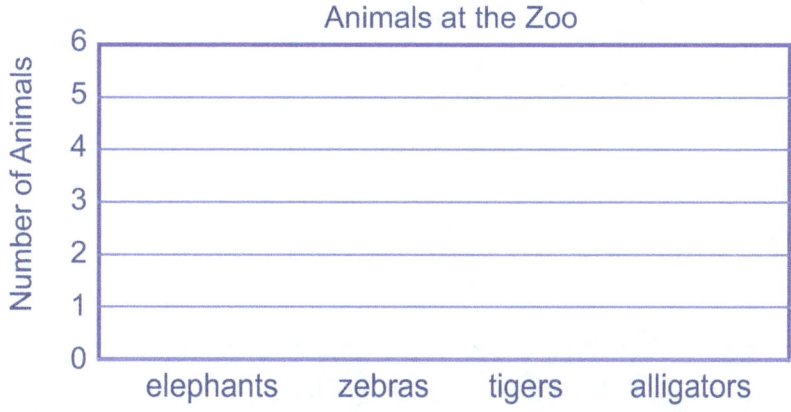

Lesson 43: Graphing Data

Independent Work - Page 1

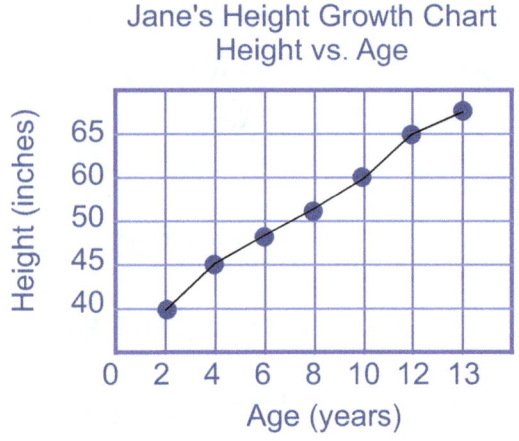

1. When Jane was 10, how tall was she?

2. When Jane was 2, how tall was she?

3. How much taller is Jane at 12 years old than at 2 years old?

4. As Jane got older did her height increase, decrease or stay the same?

5. At 13, Jane's height is about how many inches?

6. Draw a bar graph for the following: we went to the store and bought: 15 apples, 5 bananas, 6 oranges, and 10 peaches.

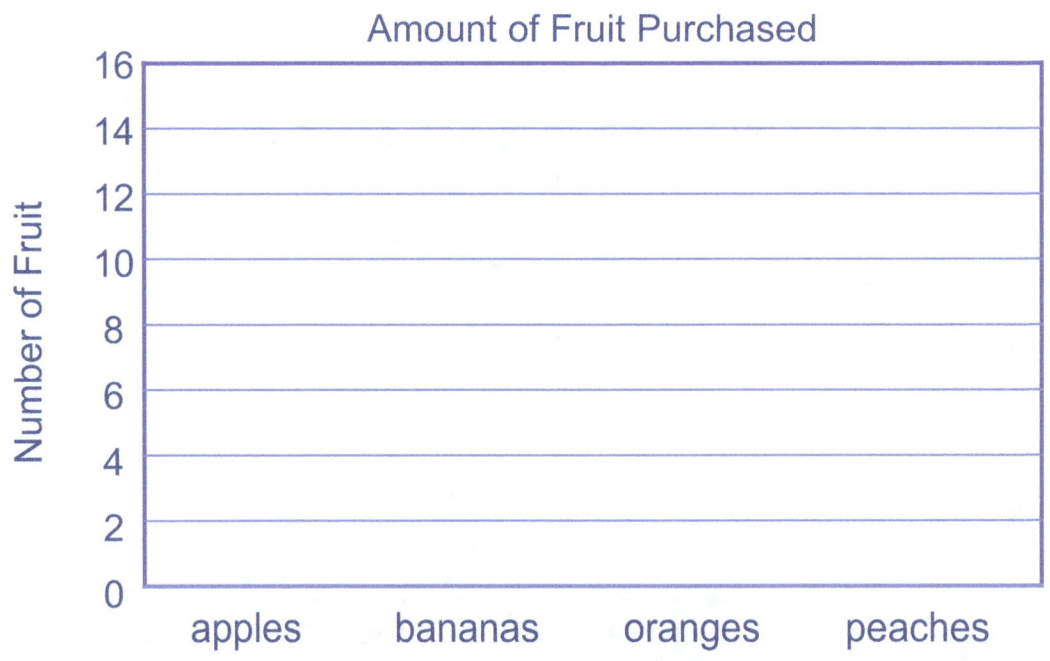

Just the Facts! Workbook

Lesson 43: Graphing Data

Independent Work - Page 2

7. What is the **volume** of the box?

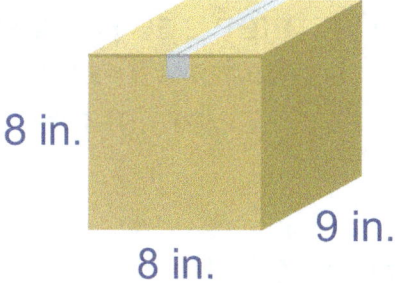

8. What is the total **volume** of the shipment if one box has the measurements below?

9.

a. What is the length of the side y? _____ ft.

b. What is the **perimeter** of the shape? _____ ft.

c. Find the unknown value to find the total area below:

$$A = (6)(8) + (5)(\underline{}) + (8)(8)$$

Lesson 44 — Patterns

We often see **patterns** in many places, which is very important, especially in science. Just like looking at a graph can help a scientist predict what will happen, so can patterns.

(44.1) Circle the balloon that comes next.

(44.2) Circle the shape that comes next.

(44.3) What letter comes next?

A A B B B A A B B B A A B B ___

(44.4) The image below is called a waveform. Continue drawing the waveform to the dotted line.

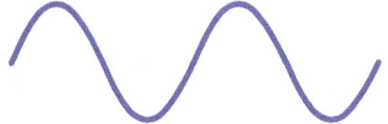

(44.5) Use a colored pencil to complete the pattern.

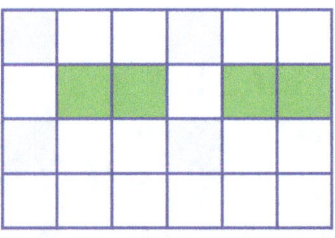

Lesson 44: Patterns

(44.6) Underline the **factors** in the equation: 4 + 2 x 2 = 8

(44.7) Underline the **product** in the equation: 6 x 4 = 24

(44.8) Underline the **sum** in the equation: 5 + 6 = 11

(44.9) Underline the **quotient** in the equation: 55 ÷ 11 = 5

(44.10) Underline the **difference** in the equation: 22 - 2 = 20

(44.11) Create a line graph, by plotting the data on the graph.

Age	Height
2	20 in.
4	42 in.
6	45 in.
8	50 in.
10	60 in.
12	62 in.

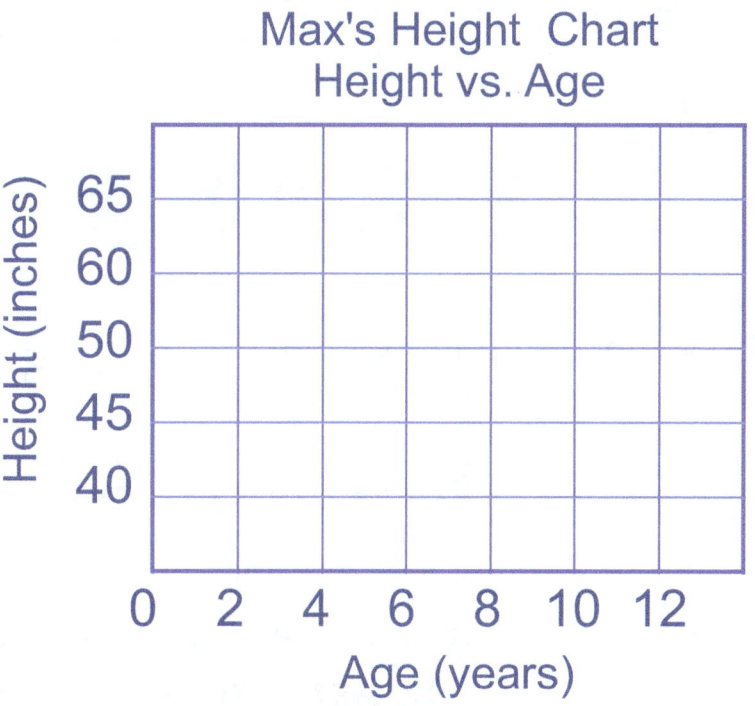

(44.12) Round 6,782 to the nearest 10. _____

(44.13) Round 6,782 to the nearest 100. _____

Lesson 44: Patterns

Independent Work - Page 1

Look at the patterns and circle the image that comes next.

1. What shape comes next?

  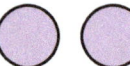

2. What ball comes next?

3. What tree comes next?

4. Which letter comes next?   A   B   C

A A A B B B C C A A B B B C ?

5. Fill in the blank with the missing shape.

6. Fill in the blank with the missing shape.

Lesson 44: Patterns

Independent Work - Page 2

Find the pattern and shade in the boxes to finish the tables.

7.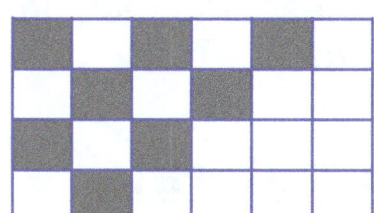

8.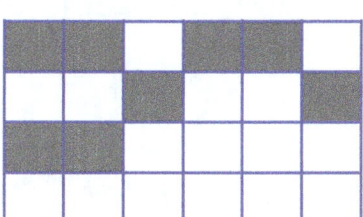

Write the fractions to show how many apples fell off the trees.

9.

10.

11.

Solve the equations below.

12. $7\overline{)56}$

13. $\frac{24}{8} =$ _____

14. $64(100) =$ _____

15. $5\overline{)45}$

16. $\frac{32}{4} =$ _____

17. $36 \times 0 =$ _____

18. Underline the **factors** in the equation: $5 + 6 \times 4 = 29$

Just the Facts! Workbook

Lesson 45: Quadrilaterals

"Quad" means four, and "lateral" means sides.
A **quadrilateral,** is a shape that has **four sides**.
We've already seen a few of these: the rectangle
and the square.

Types of Quadrilaterals

Quadrilaterals That Are Parallelograms

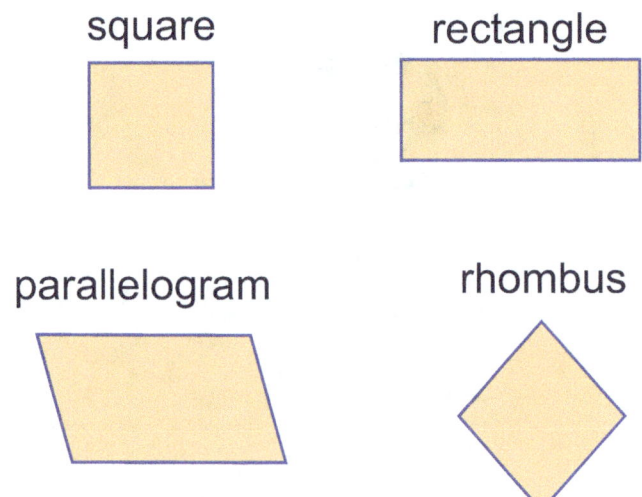

square   rectangle

parallelogram   rhombus

Quadrilaterals That Are Not Parallelograms

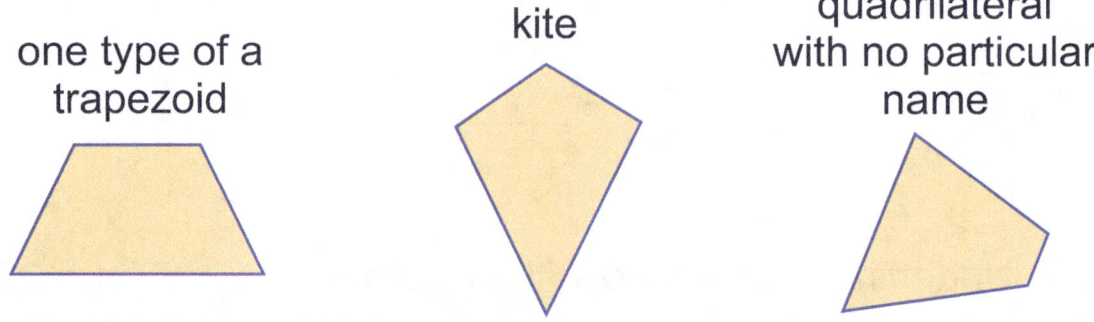

one type of a trapezoid

kite

quadrilateral with no particular name

Just the Facts! Workbook

Lesson 45: Quadrilaterals

Parallel Lines

When lines are **parallel**, they can go forever without touching.

In **parallelograms**, the sides that are *opposite* are **parallel**.

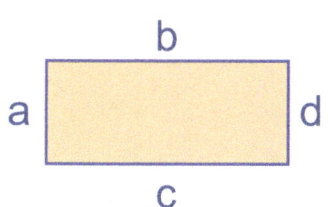 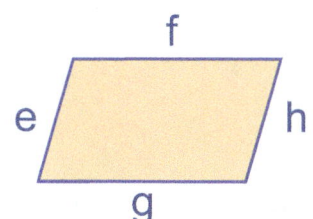 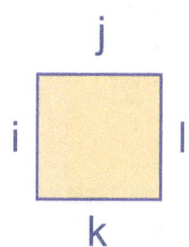

(45.1) What side is parallel to side 'g'? _____

(45.2) What side is parallel to side 'd'? _____

(45.3) What side is parallel to side 'k'? _____

The Trapezoid

The top and bottom sides of the trapezoid are parallel. The left and right sides are NOT parallel.

You can remember the name of this shape by thinking of the word "trap"; the mouse got **trapped** in the **trapezoid**.

Just the Facts! Workbook

Lesson 45: Quadrilaterals

Circle the name of the quadrilaterals.

(45.4)   parallelogram   kite   trapezoid

(45.5)   square   trapezoid   parallelogram

(45.6)   rhombus   square   kite

(45.7)   square   trapezoid   rectangle

(45.8)   trapezoid   kite   square

(45.9)   kite   trapezoid   parallelogram

(45.10)   Are the lines below parallel?   yes   no

(45.11)   Is the shape below a quadrilateral?   yes   no

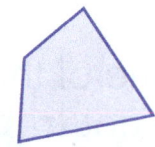

Just the Facts! Workbook

Lesson 45: Quadrilaterals

Independent Work - Page 1

1. Which shape below is **not** a quadrilateral?

   a. 　b. 　c. 　d. 　e.

2. Which quadrilateral is **not** a parallegram?

   a. 　b. 　c. 　d.

3. Which quadrilateral is a trapezoid?

   a. 　b. 　c. 　d. 　e.

4. Which quadrilateral is a parallegram?

   a. 　b. 　c. 　d. 　e.

5. Are the lines below parallel?　yes　no

   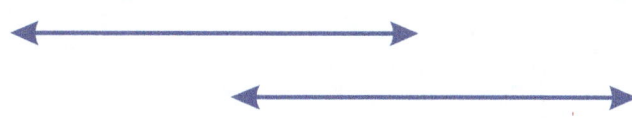

Lesson 45: Quadrilaterals

Independent Work - Page 2

Fill in the missing letter or number in the sequences below.

6. g m h p p g m h p p g m h p p ___

7. 1 0 0 ___ 1 0 0 0 1 0 0 0 1 0 0 0

8. b b R x ___ b b R x y b b R x y

---

9. A dozen donuts (12) have a mass of 120 grams.
What is the mass of each donut?
(hint: 12 x ? = 120)

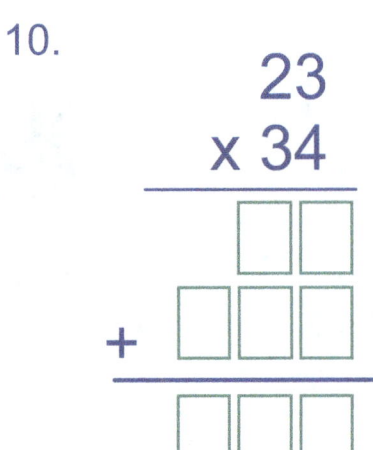

10.
```
   23
 x 34
 ─────
   ☐☐
  ☐☐☐
+ ─────
  ☐☐☐
```

11.
```
   78
 x 29
 ─────
   ☐☐
  ☐☐☐
+ ─────
  ☐☐☐
```

Just the Facts! Workbook

Lesson 1 - Skip Counting

(1.1) 2, 4, 6, 8, 10, 12, 14, 16, 18, 20
(1.2) 3, 6, 9, 12, 15, 18
(1.3) 5, 10, 15, 20

(1.4)	(1.5)	(1.6)	(1.7)
2	3	5	10
4	6	10	20
6	9	15	30
8	12	20	40
10	15	25	50
12	18	30	60
14	21	35	70
16	24	40	80
18	27	45	90
20	30	50	100

(1.8) 5,10,15
(1.9) 2,4,6,8
(1.10) 3,6,9
(1.11) 10,20,30,40,50,60
(1.12) 5,10,15,20,25,30
(1.13) 3,6,9,12
(1.14) 5,10,15+3=18
(1.15) 2,4,6,8,10+1=11
(1.16) 3,6,9,12+2=14
(1.17) 10,20,30,40+5=45
(1.18) 5,10,15,20+2=22
(1.19) 10,20,30,40,45,50,55+2=57

Independent Work

1. There are 9 fish in all.
2. There are 35 beans in all
3. There are 90 cents in all.
4. There are 25 cents in all.
5. There are 12 scoops in all.
6. There are 8 fish in all.
7. There are 20 beans in all.
8. There are 12 scoops in all.
9. There are 22 cents in all.
10. There are 31 cents in all.
11. There are 62 cents in all.
12. There are 55 cents in all.
13. There are 41 cents in all.
14. There are 28 cents in all.
15. There are 34 cents in all.
16. There are 44 cents in all.

Lesson 2 - Adding With Dots

(2.1) 5 (2.2) 4 (2.3) 3 (2.4) 2 (2.5) 10
(2.6) 7 (2.7) 6 (2.8) 9 (2.9) 8 (2.10) 11
(2.11) 10 (2.12) 8 (2.13) 12
(2.14) 15 (2.15) 10 (2.16) 13
(2.17) 19 (2.18) 23 (2.19) 34
(2.20) 24 (2.21) 21 (2.22) 43
(2.23) 33 (2.24) 62 (2.25) 52
(2.26) 40 (2.27) 92 (2.28) 103
(2.29) 25 (2.30) 6 (2.31) 10
(2.32) 8 (2.33) 30 (2.34) 12
(2.35) 20 (2.36) 19
(2.37) 13 (2.38) 22
(2.39) 34 (2.40) 38

Independent Work

1. 4, 2. 5, 3. 2, 4. 3, 5. 1
6. 8, 7. 7, 8. 6, 9. 10, 10. 9

11. 11, 12. 12, 13. 13
14. 16, 15. 12, 16. 14
17. 35, 18. 20, 19. 23

20. 8, 21. 9, 22. 12
23. 12, 24. 16, 25. 30

26. 8, 27. 19
28. 23, 29. 12
30. 28, 31. 34

Lesson 3 - Addition Double Facts

(3.1) 8 (3.2) 6 (3.3) 18
(3.4) 12 (3.5) 14 (3.6) 4
(3.7) 16 (3.8) 2 (3.9) 10

(3.14) 13 (3.15) 14 (3.16) 8
(3.17) 17 (3.18) 12 (3.19) 15
(3.20) 25 (3.21) 25 (3.22) 21

(3.10)	(3.11)	(3.12)	(3.13)
2	3	5	10
4	6	10	20
6	9	15	30
8	12	20	40
10	15	25	50
12	18	30	60
14	21	35	70
16	24	40	80
18	27	45	90
20	30	50	100

(3.23) 3 + 3 = 6
(3.24) 7 + 7 = 14
(3.25) 2 + 2 = 4
(3.26) 5 + 5 = 10
(3.27) 1 + 1 = 2
(3.28) 9 + 9 = 18
(3.29) 4 + 4 = 8
(3.30) 6 + 6 = 12
(3.31) 8 + 8 = 18
(3.32) 8 + 8 = 16
(3.33) 7 + 7 = 14
(3.34) 4 + 4 = 8
(3.35) 5 + 5 = 10
(3.36) 9 + 9 = 18
(3.37) 2 + 2 = 4
(3.38) 6 + 6 = 12
(3.39) 3 + 3 = 6

Independent Work

1. 6 + 6 = 12
2. 8 + 8 = 16
3. 2 + 2 = 4
4. 7 + 7 = 14
5. 4 + 4 = 8
6. 9 + 9 = 18
7. 5 + 5 = 10
8. 3 + 3 = 9
9. There are 16 fish in all.
10. There are 34 beans in all.
11. There are 95 cents in all.
12. There are 22 cents in all.
13. There are 19 scoops in all.
14. There are 86 cents in all.
15. 7 + 7 = 14
16. 4 + 4 = 8
17. 2 + 2 = 4
18. 8 + 8 = 16
19. 3 + 3 = 6
20. 5 + 5 = 10
21. 1 + 1 = 2
22. 9 + 9 = 18
23. 6 + 6 = 12
24. 4 + 4 = 8
25. 3 + 3 = 6
26. 9 + 9 = 18
27. 7 + 7 = 14
28. 2 + 2 = 4
29. 5 + 5 = 10
30. 8 + 8 = 16
31. 6 + 6 = 12
32. 1 + 1 = 2

Lesson 4 - Multiplication

(4.1) 4 groups of 3 dots, 4x3
(4.2) 4 groups of 2 dots, 4x2
(4.3) 6 groups of 5 dots, 5x3
(4.4) 5 groups of 4 dots, 5x4
(4.5) 2 groups of 4 dots, 2x4
(4.6) 5 groups of 3 dots, 5x3
(4.7) 4 groups of 5 beans, 4x5=20
(4.8) 3 groups of 2 fish, 3x2=6

Independent Work

1.	2.	3.	4.
2	3	5	10
4	6	10	20
6	9	15	30
8	12	20	40
10	15	25	50
12	18	30	60
14	21	35	70
16	24	40	80
18	27	45	90
20	30	50	100

5. 8
6. 6
7. 18
8. 12
9. 14
10. 4
11. 16
12. 2
13. 10
14. 4 groups of 5 dots, 4x5=20
15. 4 groups of 3 dots, 4x3=12
16. 6 groups of 2 dots, 6x2=12
17. 3 groups of 10 dots, 3x10=30
18. 5 groups of 3 dots, 5x3=15
19. 6 groups of 5 dots, 6x5=30
20. 5 groups of 5 beans, 5x5=25
21. 4 groups of 5 fish, 4x5=20
22. 6 groups of 2 eyes, 6x2=12

Lesson 5 - The Commutative Property

(5.1) 6 + 8 = 8 + 6
(5.2) 5 + 3 = 3 + 5
(5.3) 4 + 9 = 9 + 4
(5.4) 7 + 2 = 2 + 7
(5.5) 9 x 5 = 5 x 9
(5.6) 4 x 8 = 8 x 4
(5.7) 5 x 3 = 3 x 5
(5.8) 8 x 6 = 6 x 8
(5.9) There are 8 groups of 2, or 8 x 2
(5.10) There are 4 groups of 8, or 4 x 8
(5.11) There are 2 groups of 5 or 2 x 5=10
(5.12) There are 5 groups of 2 or 5 x 2=10
(5.13) Yes, they have the same number of dots.

(5.14) 4 x 5 = 5 x 4 = 20
(5.15) 7 x 5 = 5 x 7 = 35
(5.16) 4 x 3 = 3 x 4 = 12
(5.17) 6 x 2 = 2 x 6 = 12
(5.18) 4 x 10 = 10 x 4 = 40
(5.19) 6 x 3 = 3 x 6 = 18
(5.20) 7 x 2 = 2 x 7 = 14
(5.21) 7 x 10 = 10 x 7 = 70
(5.22) 2 groups of 3 dots, 2 x 3 = 6
3 groups of 2 dots, 3 x 2 = 6
(5.23) 4 groups of 5 dots, 4 x 5 = 20
5 groups of 4 dots, 5 x 4 = 20

(5.24) 5, 10, 15, 20, 25, 30, 35, 40, 45, 50
10, 20, 30, 40, 50, 60, 70, 80, 90, 100

Independent Work

1.	2.	3.	4.
2	3	5	10
4	6	10	20
6	9	15	30
8	12	20	40
10	15	25	50
12	18	30	60
14	21	35	70
16	24	40	80
18	27	45	90
20	30	50	100

5. 16
6. 10
7. 18
8. 6
9. 8
10. 4
11. 2
12. 15
13. 12
14. 4 groups of 3 dots, 4 x 3 = 12
3 groups of 4 dots, 3 x 4 = 12
15. 6 groups of 5 dots, 6 x 5 = 30
5 groups of 6 dots, 5 x 6 = 30
16. 2 groups of 7 dots, 2 x 7 = 14
7 groups of 2 dots, 7 x 2 = 14
17. 3 x10 or 10 x 3, Jack needs 30 doors.
18. 5 x 3 or 3 x5, Jack needs 15 doors.
19. 4 x 3 or 3 x 4, Jack needs 12 doors.
20. 6 x 5 or 5 x 6, Jack needs 30 doors.

Lesson 6 - Multiplying by Twos

(6.1) The total is 16.
(6.2) 8 groups of 2 trophies, 8 x 2
(6.3) 2 groups of 3 dots, 2x3 or 3+3 = 6 dots
(6.4) 2 groups of 2 dots, 2x2 or 2+2 = 4 dots
(6.5) 2 groups of 5 dots, 2x5 or 5+5 = 10 dots
(6.6) 2 groups of 4 dots, 2x4 or 4+4 = 8 dots
(6.7) 2 groups of 7 dots, 2x7 or 7+7 = 14 dots
(6.8) 2 groups of 6 dots, 2x6 or 6+6 = 12 dots
(6.9) 2x9 = 9+9 = 18
(6.10) 2x7 = 7+7 = 14
(6.11) 2x2 = 2+2 = 4
(6.12) 2x5 = 5+5 = 10
(6.13) 2x8 = 8+8 = 16
(6.14) 4 (6.18) 12
(6.15) 6 (6.19) 14
(6.16) 8 (6.20) 16
(6.17) 10 (6.21) 18
(6.22) 2 x 3 = 3+3=6
(6.23) 2 x 6 = 6+6 =12
(6.24) 2 x 7 = 7+7 =14

Independent Work

1. 2 x 4 = 4+4 = 8
2. 2 x 9 = 9+9 =18
3. 2 x 5 = 5+5 =10
4. 2 x 3 = 3+3 =6
5. 2 x 6 = 12
6. 2 x 8 = 16
7. 2 x 2 = 14
8. 2 x 7 = 14
9. 2 x 4 = 8
10. 2 x 6 = 12
11. 2 x 8 = 18

12.	8 x 2	8 + 2	8 + 8	2 x 8
13.	2 x 6	6 x 2	2 + 6	6 + 6
14.	5 x 2	5 + 5	2 x 5	2 + 5
15.	9 x 2	9 + 9	2 x 9	9 + 2

16. 2 x 9 = 18
17. 2 x 7 = 14
18. 2 x 4 = 8
19. 2 x 3 = 6

Lesson 7 - Multiplying by 3, 6 & Multiples of 10

(7.1) 3 + 3 + 3 + 3 = 12
(7.2) 5 + 5 + 5 = 15
(7.3) 5 groups of 10 cents or 5x10

(7.4) 60 (7.7) 80 (7.10) 750 (7.13) 150
(7.5) 40 (7.8) 90 (7.11) 110 (7.14) 120
(7.6) 50 (7.9) 20 (7.12) 530 (7.15) 250

(7.16) 5 x 4 = 20
(7.17) 3 x 5 = 5
(7.18) 5 x 7 = 35
(7.19) 3 x 4 =12
(7.20) 6 x 3 =18
(7.21) 10 x 3 = 30

(7.22) 10 x 5 = 5 0
(7.23) 100 x 5 = 500
(7.24) 1,000 x 5 = 5, 000
(7.25) 10,000 x 5 = 50,000

(7.26) 66 x 1000 = 66,000
(7.27) 951 x 100 = 95,100
(7.28) 6457 x 10 = 64,570
(7.29) 32 x 1,000 = 32,000
(7.30) 320 x 1,000 = 320,000

Independent Work

	number of zeros	answer
1. 100 x 7 =	2	700
2. 10 x 54 =	1	540
3. 1,000 x 30 =	3	30000
4. 10,000 x 40 =	4	400,000

5.	12 x 2	2 x 12	12 + 12	12 + 2
6.	3 x 6	6 x 3	6 + 6 + 6	6 + 3
7.	6 + 6	2 x 6	6 x 2	6 x 6
8.	4 x 3	3 + 3 + 3 + 3	3 x 4	4 + 4 + 4
9.	3 x 2	2 + 2	2 + 2 + 2	3 + 3

10. 8
11. 12
12. 6
13. 20
14. 35
15. 40
16. 9
17. 18
18. 12
19. 80
20. 40
21. 60

Lesson 8 - Multiplying by 9

(8.1) 36 (8.5) 36 (8.9) 45
(8.2) 27 (8.6) 54 (8.10) 72
(8.3) 54 (8.7) 27 (8.11) 18
(8.4) 81 (8.8) 63 (8.12) 81

Independent Work

1. 36
2. 54
3. 27
4. 63
5. 81
6. 18
7. 40
8. 900
9. 3,000
10. 540
11. 7,610
12. 2,300
13. 35
14. 15
15. 25
16. 40
17. 20
18. 30
19. 4x9= 36 dollars
20. 6+9= 15 dollars
21. 3x5= 15 books
22. 9x6= 54 dollars
23. 8x9= 72 dollars

Lesson 9 - Multiplying by 1 & 11

(9.1) 7 (9.2) 4 (9.3) 8 (9.4) 14 (9.5) 26

(9.6) 66 (9.10) 18 (9.15) 14
(9.7) 77 (9.11) 10 (9.16) 2
(9.8) 88 (9.12) 6 (9.17) 16
(9.9) 99 (9.13) 12 (9.18) 4
(9.14) 8

(9.19) 3, 6, 9, 12, 15, 18, 12
(9.20) 5, 10, 15, 20, 25, 30, 35

(9.21) 18 (9.24) 14 (9.27) 40
(9.22) 20 (9.25) 12 (9.28) 16
(9.23) 9 (9.26) 30 (9.29) 4

(9.30) 7,090 (9.31) 2,500

(9.32) 56,000 (9.33) 2,000

Independent Work

1. 8
2. 12
3. 6
4. 14
5. 18
6. 12
7. 20
8. 30
9. 40
10. 12
11. 9
12. 18
13. 27
14. 36
15. 54
16. 6,600
17. 7,4000
18. 87,500
19. 55
20. 77
21. 88
22. 66
23. 123
24. 200
25. 76
26. 983
27. 44
28. 55
29. 33
30. 64
31. 45
32. 24

33. 4x11=44
Emily has 44 soccer balls.

Lesson 10 - Multiplying by 4

(10.1) 4+4=8, 8+8=16, 4x4=16
(10.2) 6+6=12, 12+12=24, 4x6=24
(10.3) 4+4=8, 8+8=16, 4x4=16
(10.4) 4 ate (8) 32, 4x8=32
(10.5) 3+3=6, 6+6=12, 3x4=12
(10.6) 6+6=12, 12+12=24, 4x6=24
(10.7) 7+7=14, 14+14=28, 7x4=28
(10.8) 5+5=10, 10+10=20, 4x5=20
or 5,10,15,20

(10.9) 27 (10.13) 63 (10.17) 5,000
(10.10) 18 (10.14) 36 (10.18) 600
(10.11) 54 (10.15) 45 (10.19) 71,000
(10.12) 72 (10.16) 81 (10.20) 60,000

(10.21) 5x4=4x5=20
(10.22) 2+8=8+2=10
(10.23) 7x4=4x7=28
(10.24) 9+3=3+9=12

Independent Work

1. 1,200
2. 10,000
3. 22,500
4. 64,000
5. 23,000
6. 8,900
7. 81
8. 45
9. 27
10. 54
11. 36
12. 63
13. 25
14. 35
15. 30
16. 12
17. 10
18. 8
19. 6
20. 16
21. 14
22. 12
23. 28
24. 24
25. 4x6=24 beans
26. 10+5=15 beans
27. 4x8=32 beans

Just the Facts! Workbook - 243 -

Lesson 11 - The Times Table Array

(11.1) 7 x 4 = 28, double double
(11.2) 5 x 6 = 30, skip count by 5
(11.3) 9 x 6 = 54, use hands
(11.4) 81
(11.5) 110
(11.6) 9   (11.10) 52x10= 520 weeks
(11.7) 25   (11.11) 9x4= 36 pencils.
(11.8) 32   (11.12) 6x4= 24 dollars
(11.9) 76   (11.13) 25x10= 250 fingers

Independent Work

1. 54   7. 8   13. 40
2. 81   8. 36   14. 36
3. 30   9. 18   15. 66
4. 12   10. 6   16. 5,400
5. 15   11. 9   17. 48,000
6. 20   12. 35 18. 78,900

19.

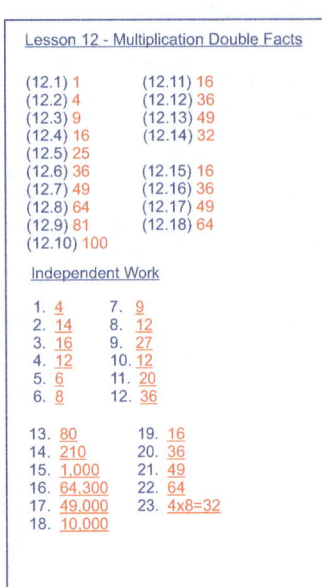

Lesson 12 - Multiplication Double Facts

(12.1) 1   (12.11) 16
(12.2) 4   (12.12) 36
(12.3) 9   (12.13) 49
(12.4) 16   (12.14) 32
(12.5) 25
(12.6) 36   (12.15) 16
(12.7) 49   (12.16) 36
(12.8) 64   (12.17) 49
(12.9) 81   (12.18) 64
(12.10) 100

Independent Work

1. 4   7. 9
2. 14   8. 12
3. 16   9. 27
4. 12   10. 12
5. 6   11. 20
6. 8   12. 36

13. 80   19. 16
14. 210   20. 36
15. 1,000 21. 49
16. 64,300 22. 64
17. 49,000 23. 4x8=32
18. 10,000

Lesson 13 - Multiplying by 3 Revisited

(13.1) 16   (13.9) 1
(13.2) 36   (13.10) 4
(13.3) 49   (13.11) 9
(13.4) 64   (13.12) 16
(13.5) 6   (13.13) 25
(13.6) 3   (13.14) 36
(13.7) 3   (13.15) 49
(13.8) 8   (13.16) 64
             (13.17) 81
             (13.18) 100

Independent Work

1. 16   5. 18   9. 7x3=21
2. 36   6. 21   10. 8x4=32
3. 49   7. 24
4. 64   8. 32

11.

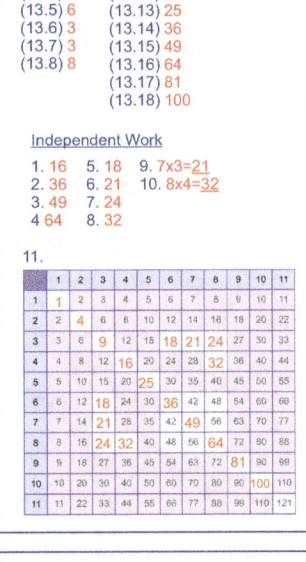

Lesson 14 - 7x6, 7x8, 8x6

(14.1) 7x7=49   (14.12) 5
(14.2) 4x4=16   (14.13) 10
(14.3) 8x8=64   (14.14) 15
(14.4) 6x6=36   (14.15) 20
(14.5) 7x6 or 6x7=42   (14.16) 25
(14.6) 6x8 or 8x6=48   (14.17) 30
(14.7) 7x8 or 8x7=56   (14.18) 35
(14.8) 3x6 or 6x3=18   (14.19) 40
(14.9) 8x3 or 3x8=24   (14.20) 45
(14.10) 4x8 or 8x4=32   (14.21) 50
(14.11) 7x3 or 3x7=21

(14.22) 81   (14.25) 12
(14.23) 72   (14.26) 14
(14.24) 63   (14.27) 16

Independent Work

1. 32   12. 32   23. 9
2. 36   13. 36   24. 12
3. 56   14. 56   25. 27
4. 24   15. 24   26. 25
5. 18   16. 18   27. 20
6. 16   17. 16   28. 48
7. 64   18. 64
8. 21   19. 21
9. 42   20. 42
10. 49 21. 49
11. 48 22. 48

Lesson 15 - Multiplying by Zero/Review

(15.1) 0   (15.4) 0
(15.2) 0   (15.5) 0
(15.3) 0   (15.6) 0

(15.7) 4 groups of 4 beans. 4x4=16
(15.8) 2 groups of 6 cents. 2x6=12
(15.9) 1 group of 4 fish. 1x4=4
(15.10) 2 groups of 6 beans. 2x6=12

(15.11) 64   (15.17) 32
(15.12) 49   (15.18) 21
(15.13) 18   (15.19) 36
(15.14) 48   (15.20) 24
(15.15) 42   (15.21) 56
(15.16) 16

Independent Work

1. 0   14. 6   27. 24
2. 11   15. 8   28. 49
3. 0   16. 18   29. 32
4. 0   17. 14   30. 64
5. 52   18. 16   31. 56
6. 0   19. 20   32. 36
7. 81   20. 40   33. 42
8. 25   21. 0   34. 48
9. 2   22. 54   35. 21
10. 4   23. 63   36. 16
11. 16 24. 0   37. 18
12. 6   25. 28
13. 10 26. 20

Lesson 16 - Factors and Products

(16.1) product   (16.4) factor
(16.2) factor   (16.5) product
(16.3) factor   (16.6) product

(16.7) factor   (16.15) 36
(16.8) product   (16.16) 64
(16.9) factor   (16.17) 49
(16.10) factor   (16.18) 16
(16.11) product   (16.19) 24
(16.12) factor   (16.20) 56
(16.13) factor   (16.21) 32
(16.14) product   (16.22) 48
                   (16.23) 24

Independent Work

1. 48   12. 33   23. 24
2. 16   13. 30   24. 28
3. 24   14. 400 25. 49
4. 32   15. 77   26. 72
5. 64   16. 0   27. 4x6=24
6. 18   17. 0
7. 49   18. 15
8. 21   19. 16
9. 36   20. 81
10. 56 21. 56
11. 42 22. 27

Lesson 17 - Commas in Large Numbers

(17.1) thousand   (17.8) 45,102
(17.2) million   (17.9) 165,500
(17.3) thousand   (17.10) 100,123,001
(17.4) million   (17.11) 65,000,000,000
(17.5) thousand
(17.6) billion
(17.7) million

(17.12) eight million, seven hundred thousand, one hundred twenty
(17.13) forty-one million, three hundred two thousand

Independent Work

1.

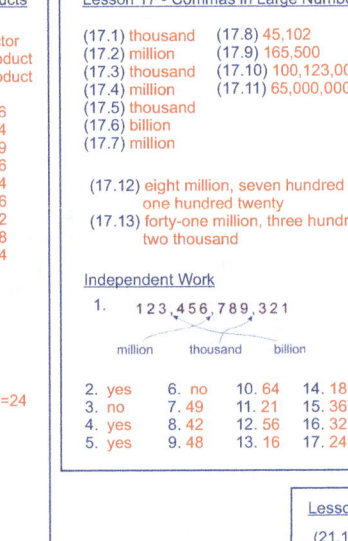

2. yes   6. no   10. 64   14. 18
3. no   7. 49   11. 21   15. 36
4. yes   8. 42   12. 56   16. 32
5. yes   9. 48   13. 16   17. 24

Lesson 18 - Place Value of Numbers

(18.1) ten thousands
(18.2) one millions
(18.3) ten thousands
(18.4) hundred million
(18.5) thousands

Independent Work

1.

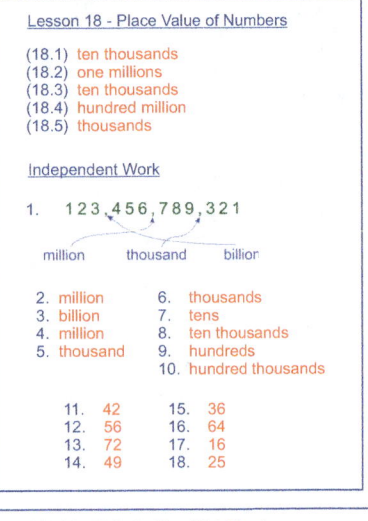

2. million   6. thousands
3. billion   7. tens
4. million   8. ten thousands
5. thousand   9. hundreds
                 10. hundred thousands

11. 42   15. 36
12. 56   16. 64
13. 72   17. 16
14. 49   18. 25

Lesson 19 - Multiplying Multiple Digits by Single Digits

(19.1) 5 x 4 = 20, do you need to carry a number?
Yes, you need to carry the 2.
4 x 4 = 16, do you need to add a number?
Yes, you need to add the 2 that was carried.
16+2= 18
The answer is: 180

(19.2) 6 x 2 = 12, do you need to carry a number?
Yes, you need to carry the 1.
3 x 2 = 6, do you need to add a number?
Yes, you need to add the 1 that was carried.
6+1= 7
5 x 2 = 10, do you need to add a number?
No, nothing was carried in the last step.
The answer is: 1072

(19.3) 66
(19.4) 64
(19.5) 36
(19.6) 72
(19.7) 56
(19.8) 75
(19.9) 212
(19.10) 738
(19.11) 300
(19.12) 924

Independent Work

1. 26   11. thousand
2. 44   12. billion
3. 93   13. million
4. 96   14. 0
5. 65   15. 0
6. 70   16. 510
7. 496   17. 125
8. 495   18. 220
9. 172   19. 2,100
10. 888 20. 2,000

Lesson 20 - More on Multiplying Multiple Digits

(20.1) 4,468   (20.5) 24   (20.10) 56
(20.2) 4,272   (20.6) 12   (20.11) 32
(20.3) 20,256 (20.7) 16   (20.12) 36
(20.4) 27,510 (20.8) 9   (20.13) 49
               (20.9) 16

(20.14) product   (20.18) factor
(20.15) product   (20.19) product
(20.16) factor   (20.20) product
(20.17) factor   (20.21) factor

Independent Work

1. 48   13. 6,396
2. 28   14. 8,944
3. 32   15. 5,256
4. 32   16. 2,256
5. 56
6. 28   17. 7x8 or 8x7=
7. 16   56 days
8. 21   18. 6x4 or 4x6 = 24
9. 49   19. 12x3=36 eggs
10. 42   20. 7x6 or 6x7=
11. 24   42 chairs
12. 56

Lesson 21 - Multiplying Multiple Digits by Two Digit Numbers

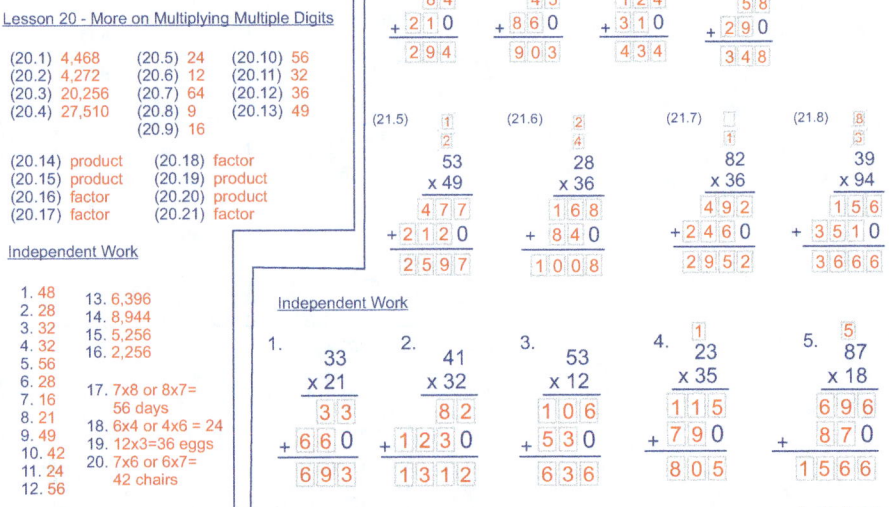

Just the Facts! Workbook

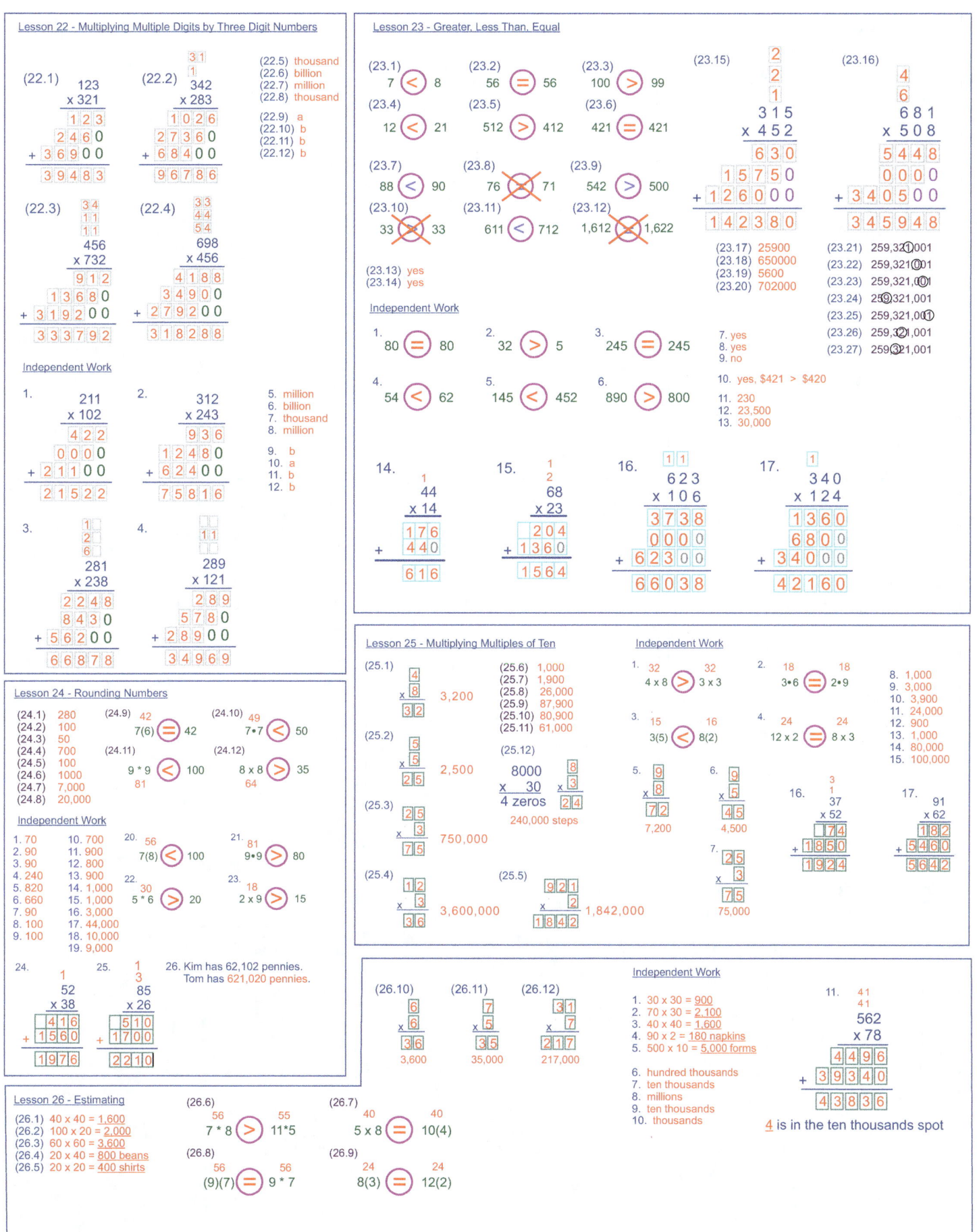

Lesson 27 - Expanding Numbers

(27.1) 25,896 = 20,000 + 5,000 + 800 + 90 + 6

(27.2) 92,341 = 90,000 + 2,000 + 300 + 40 + 1

(27.3) 168,520 = 100,000 + 60,000 + 8,000 + 500 + 20

(27.4) 6030 = 6,000 + 30

(27.5) 5,432 = 5,000 + 400 + 30 + 2

(27.6) 41,852
(27.7) 68,900
(27.8) 864,800

Independent Work

1. 13,679 = 10,000 + 3,000 + 600 + 70 + 9
2. 4,310 = 4,000 + 300 + 10
3. 231,520 = 200,000 + 30,000 + 1,000 + 500 + 20
4. 25,386 = 20,000 + 5,000 + 300 + 80 + 6
5. 684 = 600 + 80 + 4
6. 1459 = 1,000 + 400 + 50 + 9
7. 925 = 900 + 20 + 5
8. 62,739
9. 58,400

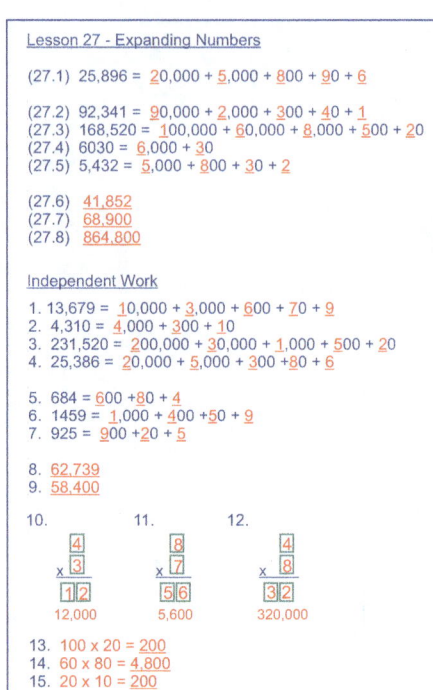

13. 100 x 20 = 200
14. 60 x 80 = 4,800
15. 20 x 10 = 200

Lesson 28 - Multiplying with the Box Method

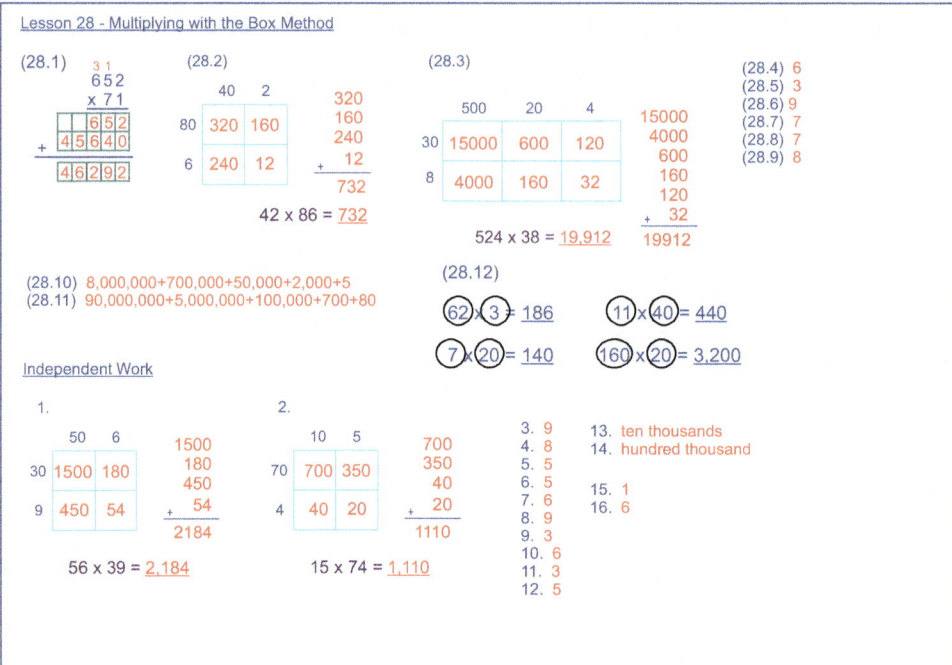

(28.4) 6
(28.5) 3
(28.6) 9
(28.7) 7
(28.8) 7
(28.9) 8

(28.10) 8,000,000+700,000+50,000+2,000+5
(28.11) 90,000,000+5,000,000+100,000+700+80

Independent Work

3. 9
4. 8
5. 5
6. 5
7. 6
8. 9
9. 3
10. 6
11. 3
12. 5

13. ten thousands
14. hundred thousand
15. 1
16. 6

Lesson 29 - The Distributive Property

(28.1) 20+15=35, 5x7=35
(28.2) 6+10=16, 2x8=16
(28.3) 42+12=54, 6x9=54

Independent Work

1. 12+8=20, 4x5=20
2. 4+16=20, 2x10=20
3. 25+30=55, 5x11=55

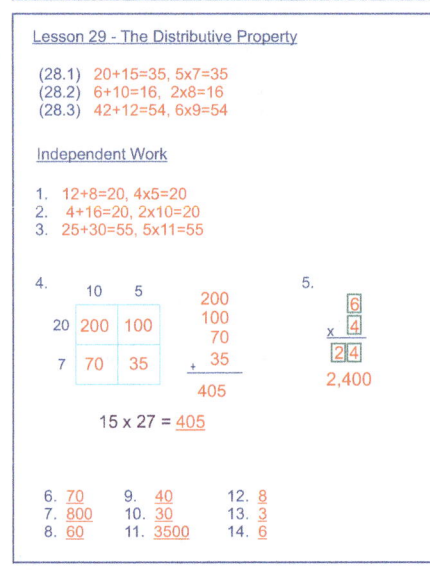

15 x 27 = 405

6. 70   9. 40   12. 8
7. 800   10. 30   13. 3
8. 60   11. 3500   14. 6

Lesson 30 - Introduction to Exponents

(30.1) 5^3 (30.5) 2x2x2x2   (30.9) 1
(30.2) 6^2 (30.6) 6x6x6   (30.10) 6
(30.3) 7^4 (30.7) 7x7x7x7   (30.11) 7x7=49
(30.4) 6^5 (30.8) 9x9   (30.12) 9x9=81

(30.13) 2^3 = 2 x 2 x 2

(30.14) 2^2 < 12   (30.15) 6^2 < 37   (30.16) 8^0 > 0
  4   36   1
  4 x 2 = 8

(30.17) 7^2 = 49   (30.18) 9^1 = 9   (30.19) 10^2 = 100

Independent Work

1. 8 x 8 x 8   6. 1   13. 6   18. 4
2. 4 x 4   7. 1   14. 5   19. 5
3. (6)(6)(6)   8. 4   15. 6   20. 7
4. 7 x 7 x 7 x 7   9. 1   16. 6   21. 8
5. 9 * 9   10. 28   17. 2   22. 3
                    11. 1   23. 365 x 10 = 3,650 days
                    12. (c)   24. (b)

Lesson 31 - Division

(31.1) Ten divided by two is equal to 5.

10 ÷ 2 = 5 $\frac{10}{2}$ = 5   2)$\overline{10}$ = 5

(31.2) Sixteen divided by four is equal to 4.

16 ÷ 4 = 4 $\frac{16}{4}$ = 4   4)$\overline{16}$ = 4

(31.3) 16 ÷ 2 = 8 $\frac{16}{2}$ = 8   2)$\overline{16}$ = 8

(31.4) 5
(31.5) 10
(31.6) 9
(31.7) 11
(31.8) 11 x 9, 9
(31.9) 9 x 6, 6
(31.10) 5 x 6, 6
(31.11) 6 x 10, 10
(31.12) 9 x 3, 3
(31.13) 6 x 7, 7
(31.14) 7 x 7, 7

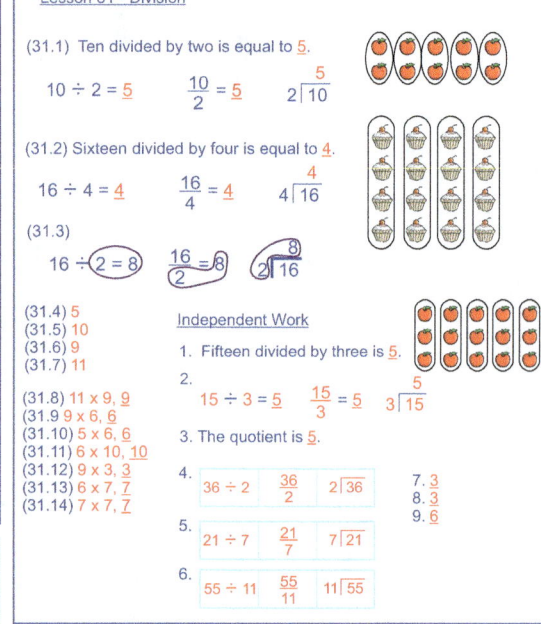

Independent Work

1. Fifteen divided by three is 5.

2. 15 ÷ 3 = 5 $\frac{15}{3}$ = 5   3)$\overline{15}$

3. The quotient is 5.

4. 36 ÷ 2 $\frac{36}{2}$ 2)$\overline{36}$ 7. 3
                                                      8. 3
                                                      9. 5

5. 21 ÷ 7 $\frac{21}{7}$ 7)$\overline{21}$

6. 55 ÷ 11 $\frac{55}{11}$ 11)$\overline{55}$

Lesson 32 - Word Problems

(32.1) 15/3 = 5 dollars

(32.2) (a)
(32.3) (b)
(32.4) (c)

(32.5) 7
(32.6) 5
(32.7) 10

(32.8) (b) & (c)
(32.9) (a) & (b)
(32.10) (b) & (c)
(32.11) (b) & (c)

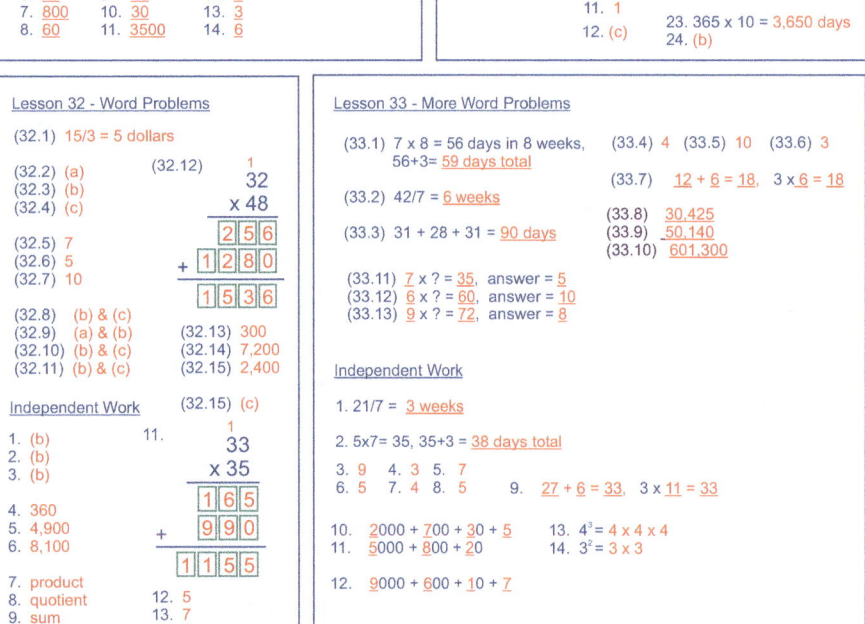

(32.13) 300
(32.14) 7,200
(32.15) 2,400
(32.15) (c)

Independent Work

1. (b)
2. (b)
3. (b)
4. 360
5. 4,900
6. 8,100
7. product
8. quotient
9. sum
10. factor

11. (shown)

12. 5
13. 7
14. 6

Lesson 33 - More Word Problems

(33.1) 7 x 8 = 56 days in 8 weeks, 56+3= 59 days total

(33.2) 42/7 = 6 weeks

(33.3) 31 + 28 + 31 = 90 days

(33.4) 4   (33.5) 10   (33.6) 3

(33.7) 12 + 6 = 18,   3 x 6 = 18

(33.8) 30,425
(33.9) 50,140
(33.10) 601,300

(33.11) 7 x ? = 35, answer = 5
(33.12) 6 x ? = 60, answer = 10
(33.13) 9 x ? = 72, answer = 8

Independent Work

1. 21/7 = 3 weeks
2. 5x7= 35, 35+3 = 38 days total
3. 9   4. 3   5. 7
6. 5   7. 4   8. 5   9. 27 + 6 = 33, 3 x 11 = 33
10. 2000 + 700 + 30 + 5   13. 4^3 = 4 x 4 x 4
11. 5000 + 800 + 20   14. 3^2 = 3 x 3
12. 9000 + 600 + 10 + 7

Lesson 34 - Fractions

(34.1) 3/4 is not shaded

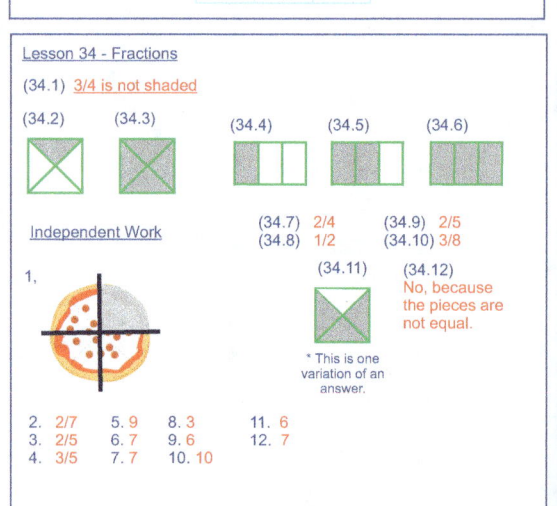

(34.7) 2/4   (34.9) 2/5
(34.8) 1/2   (34.10) 3/8

(34.11) (shown)

(34.12) No, because the pieces are not equal.

* This is one variation of an answer.

Independent Work

1. (pizza image)

2. 2/7   5. 9   8. 3   11. 6
3. 2/5   6. 7   9. 6   12. 7
4. 3/5   7. 7   10. 10

Just the Facts! Workbook

Lesson 35 - Fractions on a Number Line

(35.1)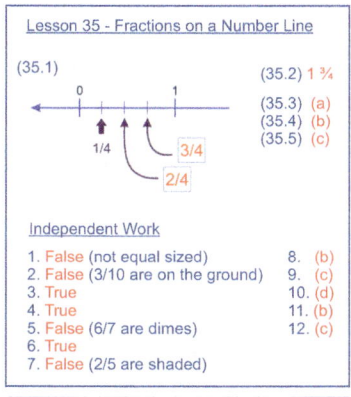
(35.2) 1 ¾
(35.3) (a)
(35.4) (b)
(35.5) (c)

Independent Work
1. False (not equal sized)
2. False (3/10 are on the ground)
3. True
4. True
5. False (6/7 are dimes)
6. True
7. False (2/5 are shaded)
8. (b)
9. (c)
10. (d)
11. (b)
12. (c)

Lesson 38 - Area of a Square & Rectangle

(38.1) 20
(38.2) 20
(38.3) 10x9=90, 90
(38.4) 5x5=25, 25
(38.5) 3/7 < 2/3
(38.6) No, just compare the numerators.
(38.7) 3/7 > 2/7
(38.8) 2:05
(38.9) 8:22
(38.10) 12:32
(38.11) clock
(38.12) clock

Independent Work
1. 42
2. 20
3. 49
4. 16
5. 100 x 65 = 6500 sq feet Need 7 bags
6. 2:40
7. 11:12
8. 1:22
9. clock
10. clock
11. 4/7 > 2/7
12. 63 9*7 > 32 8*4
13. 8/9 > 4/9
14. 8/80 = 8/2*4
15. 1 ¼ < 1 ¾
16. 20/4 = 25/5

Lesson 41 - Volume

(41.1) 3x3x3
9x3
27
(41.2) 10x10x10
100x10
1,000
(41.3) 5x5x5
25x3
75
(41.4) 2x3x1
6x1
6
6 cubes counted
(41.5) 3x4x1
12x1
12
12 cubes counted
(41.6) 3x4x2
12x2
24 cubic units

Independent Work
1. V=5x5x5
25x5
125
2. V=4x4x4
16x4
64
3. V=2x3x2
6x2
12
4. 4/8 > 1/4
5. 4/8 = 2/4
6. 2/8 = 1/4
7. 4+3+4+3+8+6 = 28
8. A=4x6=24, A = 12
A(total)= 24+12= 36

Lesson 36 - Comparing Fractions

(36.1) 3/10 < 4/10
(36.2) 7/10 > 5/10
(36.3) 9/10 > 5/10
(36.4) 5/10 > 3/10
(36.5) 7/10 > 1/4
(36.6) 4/6 > 3/10
(36.7) 5/10 = 4/8
(36.8) 4/8 = 3/6

Independent Work

1. numerator
2. product
3. denominator
4. quotient
5. factor
6. 6/10 > 2/10
7. 1/10 < 4/10
8. 4/8 < 3/4
9. 4/5 > 7/10
10. Sue ate more.
11. 6/8
12. 2 ¼ inch
13. 6
14. 6
15. 6

Lesson 39 - Perimeter of a Square & Rectangle

(39.1) 18 units
(39.2) 40 feet
(39.3) 27 inches
(39.4) 17 cm.
(39.5) 580 feet
(39.6) 6:50
(39.7) 6:53
(39.8) 7:02
(39.9) 4/9 < 3/5
(39.10) 4/8 = 3/6
(39.11) 21
33
65
876
x 359
7884
43800
+ 262800
314484
314,484
1 is in ten thousands spot.

Independent Work
1. (a) P = 22 inches, (b) A = 10 sq in.
2. (a) P = 26 yards, (b) A = 42 sq yards
3. P = 45 inches
4. P = 31 units
5. 5:40
6. 8:38
7. 7:07
8. 4/5 > 4/6
9. 1/2 = 10/20
10. 76
65
698
x 87
4886
+ 55840
60726
60,726
0 is in thousands spot.

Lesson 42 - Mass & Weight

(42.1) (a)
(42.2) (b)
(42.3) (c)
(42.4) 8x9= 72 grams
(42.5) 63/9 or 9x?=63, 7 grams each
(42.6) 128-84 = 44 grams
(42.7) 3x4x5=12x5= 60 cubic units
(42.8) 3x4x1=12
2x3x1= 6
12+6=18
18 cubic units
(42.9) A=LxW
99=9x?, or
? = 99/9 = 11
(42.10) (a) A=LxW
8x5=40
(b) P = 2(8)+2(5)
= 16+10
= 26

Independent Work
1. 98-63=35 grams
2. (c)
3. 100kg/4 = 25kg
4. 11x8=88 grams
5. V=2x4x4=8x4=32 square units
6. V=2x4x4=32, V=4x1x1=4
total volume = 32+4=36 sq units
7. (a) 56/8 or 8x?=56, ?= 7
(b) P=2(7)+2(8)= 14+16=30

Lesson 43 - Graphing Data

(43.1) more girls
(43.2) yes
(43.3) (b) no
(43.4) January
(43.5) (a) yes
(43.6) about 9 inches
*answer can vary
(43.7) March
(43.8) Friday
(43.9) Wednesday
(43.10) 10 degrees
(43.11) no
(43.12) 20
(43.13) 10
(43.14) 60
(43.15) Animals at the Zoo graph

Independent Work
6. Amount of Fruit Purchased graph
1. 60 inches
2. 40 inches
3. 65-40=25in.
4. increase
5. 67 or 68 in.
7. 8x8x9=64x9 =576 sq. in.
8. 8x11x1=88 sq in., (88)(5)=440 sq. in.
9. (a) 8+5+8= 21 ft.
(b) 6+8+2+5+6+8+8+21 =64 ft.
(c) A = (6)(8) + (5)(4) + (8)(8)

Lesson 44 - Patterns

(44.1) balloon
(44.2) green circle
(44.3) B
(44.4) wave
(44.5) checkered square
(44.6) 4 + 2 x 2 = 8
(44.7) 6 x 4 = 24
(44.8) 5 + 6 = 11
(44.9) 55 ÷ 11 = 5
(44.10) 22 - 2 = 20
(44.11) Max's Height Chart Height vs. Age
(44.12) 6,782 to 6,780
(44.13) 6,782 to 6,800

9. 5/10 or 1/2
10. 1/8
11. 9/10
12. 8
13. 9
14. 6,400
15. 9
16. 8
17. 0
18. 5 + 6 x 4 = 29

Lesson 45 - Quadrilaterals

(45.1) f
(45.2) a
(45.3) j
(45.4) trapezoid
(45.5) parallelogram
(45.6) rhombus
(45.7) rectangle
(45.8) square
(45.9) kite
(45.10) no
(45.11) yes

Independent Work
1. c
2. b
3. b
4. a
5. yes
6. g
7. o
8. y
9. 10 gms
10. 23 x 34 = 92 + 690 = 782
11. 78 x 29 = 702 + 1560 = 2262

Lesson 37 - Telling Time on an Analog Clock

(37.1) 3:00
(37.2) 8:00
(37.3) 1:00
(37.4) 3, 10, 3:10
(37.5) 10, 55, 10:55
(37.6) 1, 32, 1:32
(37.7) 1:30 (b)
(37.8) 9:15 (b)
(37.9) 3:45 (a)

Independent Work
1. 9:00
2. 2:00
3. 5:00
4. (b)
5. (c)
6. (a)
7. (c)
8. (c)
9. (a)
10. 2 ¼
11. 1 ½ or 1 ¾
12. 3/4
13. (a)
14. 67 x 29 = 603 + 1340 = 1943
15. 8
16. 5
17. 7
18. 6
19. 9
20. 6

Lesson 40 - Finding a Length Given the Area

(40.1) 5
(40.2) 6
(40.3) 1
(40.4) no, need 60 sq ft.
60 > 50
(40.5) L = 14, W = 4,
(a) P= 14+3+9+4+5+7 = 42
(b) A = 14 x 4 = 56 sq units
(40.6) 9,166 to 9,170
(40.7) 3,482 to 3,500
(40.8) 11,452 to 11,500
(40.9) 2,000
(40.10) 1,800
(40.11) 6,300
(40.12) 270
(40.13) 5,600
(40.14) 56 + 48 = 104
7 + 6 = 13
8 x 13 = 104
(40.15) 12:27
(40.16) 5/15
(40.17) 10/15

Independent Work
1. 8 x 8, 8 feet
2. 10 x 10, 10 feet
3. 88 sq ft, no
4. 2,400
5. 7,200
6. 3,200
7. 36,000
8. 420
9. 32 + 36 = 68
8 + 9 = 17
4 x 17 = 68
10. 9:33
11. 5/12
12. 7/12

Just the Facts! Workbook - 247 -

www.ingramcontent.com/pod-product-compliance
Lightning Source LLC
Chambersburg PA
CBHW082115230426
43671CB00015B/2709